EVOLUTIONS

Oren Harman is the author of *The Man Who Invented the Chromosome* and *The Price of Altruism*, which won the *Los Angeles Times* Book Prize for best book in Science and Technology. He has a doctorate from Oxford University, is the Chair of the Graduate Program in Science Technology and Society, and a Professor of the History of Science at Bar Ilan University.

ALSO BY OREN HARMAN

The Autobiography of Charles Darwin
(scientific ed., Hebrew translation)

Dreamers, Visionaries,
and Revolutionaries in the Life Sciences

Handbook of the Historiography of Biology

The Man Who Invented the Chromosome:
A Life of Cyril Darlington

Outsider Scientists: Routes to Innovation in Biology

The Price of Altruism:
George Price and the Search for the Origins of Kindness

Rebels, Mavericks, and Heretics in Biology

EVOLUTIONS

FIFTEEN MYTHS THAT EXPLAIN OUR WORLD

OREN HARMAN

An Apollo Book

This is an Apollo book first published in the UK in 2018 by Head of Zeus Ltd
This paperback edition published in the UK in 2019 by Apollo

9 7 5 3 1 2 4 6 8

A catalogue record for this book is available from the British Library.

ISBN (PB): 9781788547581
ISBN (E): 9781788547598

Designed by Jonathan D. Lippincott
Illustrations by Ofra Kobliner

Cover design: Rodrigo Corral and Sungpyo Hong
Illustration: Ascidiacea from Ernst Haeckel's *Kunstformen der Natur*;
courtesy of Craig Ellenwood / Alamy Stock Photo

Printed and bound in Great Britain by
CPI Group (UK) Ltd, Croydon CR0 4YY

Head of Zeus Ltd
First Floor East
5–8 Hardwick Street
London EC1R 4RG

WWW.HEADOFZEUS.COM

These stories are dedicated to my children,

Shaizee and Abie,

and to my love, Yaeli

CONTENTS

EVOLUTIONS

INTRODUCTION

The great creator Viracocha rose from Lake Titicaca during the time of darkness to bring forth light. First he made the Sun, the Moon, and the stars. Then, breathing into the stones, he made the humans. But the humans were brainless giants, and soon they displeased him. So he ruined the world he had created with a flood, and made the humans once more from smaller stones. Eventually, walking on the waters, he disappeared across the Pacific Ocean and never returned again. But the Inca believed that he sometimes walked the Earth disguised as a beggar, teaching the people the secrets of civilization and weeping over the plight of mankind.

What does this mean? It depends who you ask. In the eyes of the philosopher, myths are allegories for philosophical truths; in the eyes of the historian, they are perversions of historical truths. To the psychologist myths reflect our deepest fears; to the moralist they provide a compass, and to the poet, inspiration. Myths are stories about a distant past or an imagined future, shadowing our existence like intimate, mysterious companions. They orient us in the universe and provide a kind of comfort. But myths also summon truths beyond our jurisdiction: about the nature of matter, and time, and the forces. About how we came to be, and why we can or cannot hope, and where we might be headed. However much we try to capture their meanings, we can only grasp at them fleetingly, like children lunging after butterflies.

Perhaps the literary critic Northrop Frye came closest to capturing their essence when he said that myths describe not what happened but what happens. Frye probably cribbed this from the pagan thinker Sallustius, the author of *On the Gods and the Cosmos*, who wrote of myths: "These things were never, and are always." And Sallustius himself would have read Plato's *Phaedrus*, where the Greek philosopher calls myths "plain tall tales" while confessing that unique fictions may express general truths. Whether myths are conveyed esoterically or symbolically, some believe that they begin with a kernel of fact about the cosmos and the relations within it. But do facts ever really exist outside of how we make them? Is there such a thing as "reality," unambiguously presented to human consciousness and then described by language directly? Perhaps our concepts organize our world, before we perceive its so-called certainties. And perhaps these concepts are intimately tied to who we are.

And so, whether you are a philosopher, a moralist, a historian, a psychologist, or a poet, or just someone who loves to read myths, here is a useful way to think of things: myths are expressions of existential conundrums, creatures of our lonely, searching minds. And since our minds have always both imagined infinity and lived, uneasily, with the surety and sadness of death, throughout the ages the themes of myth have remained strikingly unchanging. Myths are humankind's stories about what we all feel in our guts is fundamental to our humanity but know with our brains can never truly be plumbed. Motherhood, Freedom, Death and Immortality; Memory and Jealousy and Solitude and Sacrifice; Birth and Rebirth, Truth, Love, Hubris, Fate—these are the realms of our indispensable mythologies.

Mythological themes may be universal, but they did not always mean the same thing to different peoples. To the ancient Egyptians, and later to the Christians, resurrection meant redemption, originally related perhaps to fertility rites of the yearly cycle of vegetation. But to modern minds who devised villains from Dracula to Batman's Joker, those who return often seek revenge and destruction. Despite the changing mores, the unannounced feeling that the themes of myths can never truly be penetrated is common to the ancient Greek, the medieval Mogul, the Bushman, the Algonquian, the Laplander, and you and me: Why are we here? How did we get here? What is the meaning of the hidden forces shaping our certain deaths and unpredictable lives?

As far as we know, rocks and trees and toads do not ponder such questions, and so, whatever our beliefs, we take myths as signs of our human distinctiveness. Certainly, they go back to our beginnings: most probably myths were not foreign to the hands that drew haunting images of animals in mountain caverns, such as the Chauvet Cave in southern France, tens of thousands of years ago. However we construe them, myths are not lies, nor are they pure inventions. Instead they are an odyssey, powerfully grounded in our collective human experience. Like life, which they set out to illuminate, they are both tangible and abstract, actual and imagined. But in the final reckoning, they do not provide definitive answers because the questions they raise have no definitive answers. Digging deeper than any other form of thought, myths represent humankind's gloriously futile quest for existential understanding.

The matter of myth is therefore more profound than the material of morality, which in comparison seems parochial, a child and perfect captive of its age. Myth is more defining than the functional fabric of religion, which comes from the Latin *ligare*, meaning "to bind." As for the objective kingdom of fact, it has been well said that mythology is the penultimate truth—penultimate because the ultimate cannot be put into words. It seems Plutarch got it right, then, when he wrote that truth and myth have the same relationship as the sun to the rainbow, which dissipates light into iridescent variety. Myths have been spoken in different tongues throughout the ages, but their root and concerns remain universal.

As all peoples do and always have done, the ancient Greeks fashioned their myths in the idiom at their disposal. Even their greatest rebels could speak only in the tongue of their time. Take, for example, Empedocles.

Born in Acragas, Sicily, Empedocles lived sometime between 493 and 423 B.C., and was a magician, a minstrel, a politician, a doctor, a philosopher, a tragedian, a charlatan, a prophet, or perhaps all of the above. He was teacher to Gorgias, pupil of Parmenides, contemporary of Zeno, follower of Xenophanes, and, like Plato, a student of Pythagoras accused of *logoklopia*, or the stealing of ideas. Empedocles, it is said, allayed a young man's murderous rage with a soothing melody on the lyre; wrote forty-three tragedies; checked a disease-carrying wind; and saved the women of the town from barrenness by blocking a cleft in the mountains. As if this were not enough, he also stopped a storm cloud from overwhelming the people of Acragas; revived a woman who had been for thirty days without sign of breathing or pulse; and cleared the acropolis of a plague caused by an evil stench from the river by digging channels from two neighboring rivers at his own expense. So wonderfully mystifying was this ancient sage of the Greek world that no one could agree on how it was he died: Demetrius of Troezen claimed he hanged himself. Favorinus wrote of a fall from a carriage and a broken thigh that never healed. Telauges had it that as an old man Empedocles had lost his balance on board ship and drowned. Heraclides Ponticus, and after him Horace and Ovid and Lucian, all great poets of the ancient world, provided the most dramatic death of all: Empedocles, according to them, dove headfirst into the gushing mouth of Mount Etna in order to confirm the report that he had become a god.

God or mortal, wondrous physician or quack, one thing about Empedocles is certain. It is he to whom we owe the four eternal elements that were to order man's understanding of nature for over two thousand years following his mysterious disappearance. "Hear, first, the four roots of all things," he proclaimed in his *Physics*: "shining Zeus, life-giving Hera, Aidoneus, and Nestis, who with her tears supplies mortal springs with water." Aghast at the untruths of the Hesiodic storytellers of his time, Empedocles desperately sought firmer ground upon which to build man's worldly knowledge. A *logos* was a demonstrable proposition, not a *mythos* or a lie. It was certainty he was after, not a filigree of fancy, nor the extraordinary, nor the unimaginable imagined. Who were Zeus, Nestis, Aidoneus, and Hera? They were Fire, Water, Earth, and the boundless height of Air. And everything the eye could see was made of them.

But why fashion the elements as gods? Why disguise the *logoi* as *mythoi* if it was the awakening from juvenile stupor that Empedocles sought above all else? A crusader and a contrarian, Empedocles wasn't trying to rescue history from pure invention. Here was a challenge aimed at the very core of ancient belief: mythical figures are mere code names for the unexplained powers of nature. Zeus, Hera, Aidoneus, and Nestis aren't the fire, air, earth, and water perceptible to the senses, literally. They are abstract powers signifying the indestructible and eternal nature of the elements we call reality. *Your gods are creations!* Empedocles shouted, just before hanging himself, or falling out of the carriage, or into the ocean, or jumping headfirst, in a fit of drunken hubris, into the roaring, blazing mouth of Mount Etna, disappearing forever to the world.

This was his legacy. But then came the moderns.

If the gods were mythological inventions that allowed the Greeks to plunge into greater depths of understanding, modern humans could do better, or so they thought. Ancient skeptics needed their deities, but modern skeptics could do away with God entirely. Myths were suddenly rendered superfluous, a new concreteness having come to the world of the steam engine, the train, the telegraph, and the electric light. Science, modernity's Promethean incarnation, was rapidly wresting the secrets from the heavens.

History would chronicle the great achievements. An English magus single-handedly fashioned a new cosmos made of Matter, Force, Gravity, Acceleration. After Newton, the Russian chemist Mendeleev ordered the elements, unallegorically: Bromine, Calcium, Hydrogen, Zinc. Mendel, a Moravian monk with a passion for peas, discovered the laws of heredity linking generation to generation. Independently, squabbling, Pasteur and Koch disciplined disease. Darwin, the bearded man from Down House, fitted all life-forms on a tree of descent with modification. And Rutherford, a hefty New Zealander with a quarter smile permanently fixed on his face, split the indivisible atom—a feat Democritus and his friends had thought impossible. Inexorably, science was disrobing the mythic world of its wonder. Empedocles may have needed his gods to do away with them, but modern man suffered no such affliction. Finally, he could read Nature's book unalloyed.

Since the dawn of Rutherford's and Mendel's century, science has proven a muscular method for getting to know the world and ourselves. It has become the magisterial language of our age, our chosen idiom. Beyond physics and chemistry, biology—the most lawless of all sciences—has fashioned itself the Ur-language of the times, bringing us closer than we ever were to understanding the very natures of those who once drew on cave walls and produced our first mythologies. The artists of Chauvet depicted galloping bison and menacing leopards, and the ancient Greeks a tempestuous Hera and a cunning Zeus, but we now invoke genes and natural selection to explain fear and desire, jealousy and hope. Instead of weeping with Virgil's Dido as Aeneas's ship pulls away from the Carthaginian bay, we pretend to capture something important by speaking of emotion with words such as "acetylcholine" and "serotonin," "dopamine" and "oxytocin," and we treat ourselves with drugs that help modulate our behaviors. The titles of our books tell the story: *The Biology of Trust*, *The Evolution of Morality*, *The Anatomy of Violence*, *The Science of a Meaningful Life*. The logic of natural selection and the intricacies of genetics and development have become our modern-day tongue for discussing motherhood and memory, the origin of morals, and the meaning of death.

We have come a long way. Since the age of Francis Bacon a techno-scientific revolution has transformed our world—how we eat, and travel, and make war and love. Predictions based on esoteric abstract mathematics have been validated and sold for billions. Huge industries have risen, providing energy for the masses and protection for our bodies and ideas. Even our greatest enigmas seem primed to be unraveled. Many scientists today celebrate the telling connections between particular areas of the brain and states of human consciousness. Observing such insults as anoxia and anesthesia, uncovering the molecules involved in depression and elation, few doubt that consciousness emerges from specific mechanisms rooted in particular brain locations and consistent with the principles of biology. Consciousness is a hard problem, but despite centuries of befuddlement we don't view it as untouchable. The future will uncover more about how the brain creates the mind. We are the makers of our brave new world.

And yet the "how" of consciousness is one thing and the "what" another. Even if we could divine the modern alchemy, the precise ratio of oxytocin and acetylcholine, dopamine and serotonin, necessary to bring about happiness in all humans under all conditions—a rather incredible assumption—knowing its chemical trigger can tell us little about what happiness *means*. We may uncover neural correlates for our emotions, but we remain ignorant of what they are at the level of experience. The brain creates the feeling of self, of this we're confident. But what is the essence of this "I"? No one can say.

Perhaps that should not surprise us. Science pretends to be a replacement for mythology, but in reality it is driven by the same hunger for understanding that brought us the gods and the afterlife, souls and creation myths, and it too is shaped by tales. Science is a form of competitive storytelling: it gives names to things, and produces narratives based on a method that has undergone impressive refinement. In this sense it is very special indeed. But even great stories are still stories—even stories that help fly airplanes and rid us of disease. In a scientific age, we should expect a mystery like consciousness to be couched in the language of physical processes. Ever since the seventeenth century, importing metaphors from different fields, we have fashioned the brain as a system of hydraulics, an intricate clock, a telegraph switchboard, a neural network, a quantum computer; we have spoken of levers being pulled and signals being processed and the emergence of nonlinear dynamics. We spin the best tales we can conjure, and our tales, just like our cultures, change with the times.

Our theories are expressions not only of the world around us but also of the different ways in which we wish to see ourselves in our world—the two are constitutive of each other and ultimately inseparable. Just as the coal sketches were to Cro-Magnon, and the Olympians were to the Greeks, the sciences serve as both our guide and our reflection. We might not think of ourselves as part of a continuum, but every age without exception frames its profound existential puzzles in the terms of its greatest achievement. We are the children of Empedocles, after all.

Despite science's willful break with mythology, and notwithstanding its marvelous achievements, let's be brave enough to ask ourselves: Are we any better off than the Greeks were when it comes to wrapping our minds and our hearts around our existential puzzles? Has the knowledge of the inflating universe gotten us closer to understanding Fate? Has the correlation between oxytocin levels in the blood and "prosociality" unmasked the essence of Motherhood? Have the shadows of Jealousy or Love or Sacrifice been further illuminated by the understanding that emotions must have conferred an advantage in evolution? Even if science will one day be able to read our thoughts by mapping interactions between the nerve cells in our brain, will we be able to say—and believe—that we have gained a deeper understanding of experience?

Science today is our safest path to knowledge, and wedded to technology it continues to amaze. Just ask anyone who has ever surfed the Web on a transatlantic flight, or popped Ritalin before an interview, or glimpsed an unborn baby through an ultrasound, or sat under an LED with a genetically modified apple to read a book at night. Science is our best path to providing answers to the type of questions that have solutions. There, incontrovertibly, it has no peers.

And yet, too frequently control is confused with understanding, by scientists and the rest of us. Too often what is natural is taken for what is right, complexity simplified into grotesque caricature, and modesty thrown to the side. Most disconcerting is something fundamental: we champion atheist horsemen, making heroes of those who tell us that the only mysteries worth revealing will succumb one day to our inquiries. This is a grave misjudgment. For there are many worthy mysteries that are not on a path to resolution. Science gives us knowledge but will not alone deliver wisdom. If we fail to define its promise, and what lies beyond its reach, our guide and reflection is in danger of becoming a hollow pledge.

Ultimately, we can do no better than to disguise our search for meaning in the language of our time, just as Empedocles did before us. Just like the cavemen and the ancient Greeks and the medieval Moguls, the Bushmen and Algonquians and Laplanders, we continue our odyssey in the human-experienced world of unlimited imagination and undeniable demise. All cultures have always known that there are places so deep not even knowledge can penetrate. "Natural science will never discover for us the inside of things . . . that which is not appearances," Immanuel Kant proclaimed in the age of Joseph Priestley and William Herschel and Edmund Halley. "Even if all possible scientific questions be answered," Ludwig Wittgenstein professed in his *Tractatus*, a century and a half later, "the problems of life have still not been touched at all. But of course there is then no question left, and just this is the answer." Both men knew what the great mythographers have always intuited: unlike in science, knowledge is not a relevant commodity in the unforgiving realm of myth.

The ancients recognized: the truths of myths are beyond our authority. They are even beyond what we can know. Yet science pretends to achieve control through knowledge, to replace submission to providence with the mastery of manipulation. It promises complete understanding, and in too many hands, complete deliverance. Synthetic biology, drug design, eco-engineering, artificial intelligence: all will come to the rescue of humanity. And when all the gaps in our knowledge are closed, and all possible meaningful queries answered, the good will surely follow.

The pledge is enticing, but it leaves many spirits wanting. Ending world hunger and achieving longer life spans are goals to be championed with all our might, but, along with "perfect" babies, they should not be confused with happiness. Intergalactic travel may be coming, but it will not reveal to us the meaning of fate, nor will a cure for cancer, however wonderful, bring existential deliverance. For good reason, looking around our world of iPhones and antibiotics, people wonder: *Is this really all there is?*

Perhaps that is why many of us continue, despite modernity, to dress our existential puzzles up just like Empedocles did, but with one important twist. We may not always rediscover Thor and Apollo, but we intuit that oxytocin and serotonin won't be able to fill their sandals. We somehow know that there are questions left untouched by science, and that these questions matter, even matter most. And so we fabricate upgraded versions of the old heroes: Spider-Man and Batman, Wonder Woman and Mr. Incredible. We invent imaginary places, like the Seven Kingdoms of Westeros and the Matrix and Narnia, and return, perennially, to *Harry Potter* and *Star Wars.* Searching for unknowable meaning, we find our way to poetry and literature, to music and art and the body, to nature and God and philosophy—anywhere but science. For science promises to cast aside mythology.

But does it have to, really?

In the myths that follow, the eternal contemplation on the wonders of life will be retold and reinvented in the language of our times. Here astronomy, physics, geochemistry, and biology will be our chosen idioms. Evolutionary theory in particular uses the latest knowledge from genetics, development, chemistry, neuroscience, geology, paleontology, and linguistics to crack the mystery of life's beginnings and the sweeping saga of its unfolding. It hopes to uncover how life started, when the first cell was born, and why cells came together to form bodies in the first place. It offers theories about the great transformations of life on earth: the origins of photosynthesis and sex and vision and flight and language and consciousness. It tells us how we got here and how we came to be.

Too often science is presented as definitive, and its lessons from history forgotten. For centuries the Ptolemaic astronomers assumed the existence of heavenly epicycles, and the Neoplatonists bowed to a Great Chain of Being. Aristotle was certain that women had fewer teeth and colder blood than men, Galen that the heart stoked our bodies, and Descartes that "I think, therefore I am." We may no longer believe these stories, nor speak of phlogiston, pangenesis, or the ether. But we should make a point of reminding ourselves that what we call "true" today will likely be "false" tomorrow; that the fate of some of the choicest theories of our greatest minds will be no different from those discarded theories of Aristotle, Galen, and Descartes. Despite our trust in the enterprise, science is not a march toward truth in some absolute sense, but rather a way to erase the "truths" of yesterday as we recast our present knowledge. Astronomical and evolutionary theories, in particular, deal with the deep past, and we must concede that many of them are highly speculative. The astronomer Edwin Hubble reminded his day that "the scientist explains the world by successive approximations," and he was right. But the nineteenth-century utopian writer and evolutionist Samuel Butler may have come closer to the heart of things when he cautioned, "Science, after all, is only an expression of our ignorance of our own ignorance."

And yet, as speculative and approximate as our theories are on the origin of matter, the galaxies, and human feelings and consciousness, they remain our best guesses for how things unfolded in our universe. Scientific theories represent the most honest attempt of our age to explain our greatest mysteries. In this sense they are no different from the gods of Olympus, though they sometimes trick us into believing that we have control over what lies beyond our reach. Empedocles may have been a skeptic, but many Greeks believed their myths just as many of us believe our science.

And just as the Olympians were for the Greeks, so too will the forces and elements and molecules and organisms in the myths that follow be our prisms for filtering existential conundrums.

There are scientists and writers who tell it like it is, faithfully juxtaposing what we know against what remains beyond our grasp, but *Evolutions* will take a different path. Here, what we think we know will serve to highlight the limits of our understanding—the latest cutting-edge knowledge marshaled, against the tide, to celebrate the perennial mystery of what resides within. Science has neglected mythology by pretending to render it superfluous. But this robs our modern-day language of an important duty, one that all languages of all times have faithfully carried out. For alongside answering questions that do have answers, science can also, surprisingly, help us live more comfortably with the uncertainty of wonder, reminding us that outside the realm of knowledge the highest we can reach is for penultimate truths. This, then, will be the mandate of *Evolutions*: through storytelling, to reclaim an age-old task for our flamboyant modern tongue.

And so here we encounter the big bang and the multiverse, and through them take a newfound look at Fate. We watch the young moon being flung by a giant meteor off the earth, and contemplate the meaning of Motherhood. The promiscuous shuffling of genes between our most ancient ancestors becomes a challenge to our modern notions of Love; the first act of symbiosis exposes the Janus faces of

Freedom; the birth of the eye on a trilobite introduces the world to Jealousy; the Memory of an octopus uncovers the paradox of consciousness; the birth of language narrates our struggle with Truth. Throughout the stories, a decisive development in the history of life and the universe will be linked to an existential theme, inviting the reader to consider that theme under the beam of a new light. In the final chapter, I return full circle to the question of who we are and why we need our myths in the first place.

A brief word on narrative: Despite attempts to avoid doing so, scientists routinely imbue their subjects with agency, be they molecules, organisms, or even physical forces. Together with their equally common use of metaphor and allegory, this represents a limitation of language, a genuine philosophical paradox, and a source of creativity. Scientists speak of energy, simultaneity, bonding, and altruism. They describe electrons "jumping," stars "shooting," flowers "opening up," pigeons "cooing," and monkeys falling in love. All such concepts have a meaning in our own lives. It is difficult, and perhaps impossible, to rid scientific definitions of the human nuances that are contained in them. Science is a language more mythic than we care to admit.

Many of the myths in *Evolutions* will be told in the voice of an all-knowing narrator, serving as a proxy for science, trying and sometimes failing to speak properly disinterested, non-teleological "scientese." Others will be told in the voice of a skeptic, drawing back from the promises of science and questioning their truth. And in keeping with the attempt to loyally reflect the enterprise, sometimes the evolutionary and cosmological figures themselves will narrate the story, using science, alongside their feelings, as both their guide and reflection. The earth will thus present a cosmologically inflected view of Motherhood; an amoeba will help us look at Pride in an unfamiliar way; a trilobite will mourn Jealousy to another; an ungulate-turned-whale will expound on the meaning of Sacrifice; and the loneliness of consciousness will emerge from the Memory of an octopus. Love, too, will be given its due by early microorganisms: we think of it as an emotion, but perhaps there are different ways to define it, with evolution as our guide.

At the end of the book, a section called "Illuminations" will invite readers to engage the evidence directly by delving more deeply into the scientific literature. The sources cited will include recommendations of popular accounts as well as classical ones; technical papers alongside textbooks, websites, and films; literary and historical allusions, often idiosyncratic. Each relevant illumination can be read directly after reading each myth, or consulted en bloc after you've read them all—it's your choice; "Illuminations" as a whole is meant to provide context and explanation, clarifying the scientific ideas as well as the relevant weight of controversial knowledge. Like all languages in the past that were used for writing mythology, many of the current scientific theories are provisional, some even highly contested. It was said of Newton that he died as he was born: a virgin.

But it is also true that every scientific theory is born refuted, if history is any kind of counselor.

Saul Bellow once said that science has been a housecleaning of belief. Whether or not he is right, science doesn't have to be a housecleaning of mythology. Some things, after all, are forever deeper than anything we can know. These stories are a resounding celebration of science, but they are also a corrective to its immodesty. "The roaring of lions," William Blake wrote, "the howling of wolves, the raging of the stormy sea, and the destructive sword, are portions of eternity too great for the eye of man." We remember this when we glance at the paintings at Chauvet, or read about Zeus and Apollo—even when we settle into our comfortable theater chairs to watch *Star Wars* or *Harry Potter*. But we forget it when we put our trust blindly in science, or else reject it entirely.

Here we walk the middle road, led by new kinds of heroes: Gravity and Angular Momentum, the Sun and Earth and Moon, Oxygen and Mitochondria, Bacteria and Ribozymes, slime molds, trilobites, early whales, pterosaurs, octopuses, and Homo erectus, all the way to humans.

On one plane the tales will carry us through the unfolding history of the universe and life on Planet Earth, on the other through the perennial themes of human mythology. Philosophers may recognize allegories for philosophical truths, psychologists our fears and desires, moralists our compass, and future historians—who knows?—perhaps telling perversions of how we once saw ourselves in our world. But I hope that all readers, insofar as you are still fully human, will recognize an age-old journey, an ancient and meaningful quest.

"These things were never, and are always."

FATE

THE BIRTH
OF THE UNIVERSE

1

Pythagoras called it "the All."

When it came to be, it already contained all that was and all that will be, all the matter and all the energy, all the stars and planets and galaxies, the budding leaves and broken hearts. Every drop that would ever evaporate, or fall silently upon a rock, all were there from the beginning. The beginning of time gave birth to all future time, to philosophy and mathematics. Believing this truth, men have desperately sought to flee the consequence.

Nearly 14 billion years ago was when "the All" began; if you had blinked, you would have missed it.

We call it the Universe.

2

But the Universe was not always the Universe. In the beginning it was merely the World, at least according to the early humans.

The Babylonians thought that the vaults of the Earth and the Heavens were wrought from the ribs of Tiamat, torn to pieces by her son Marduk, her teary eyes becoming the Tigris and the Euphrates, her tail the Milky Way.

The Norse, to the contrary, had it all begin with a giant created when fire and ice met in the abyss of Ginnungagap. As Ymir was suckled by the cow Auðhumla, his sweat gave rise to more giants, spontaneously, and these to still more descendants, including Odin. It was Odin's two brothers, Vili and Ve, who slew their progenitor's progenitor, as told in the *Song of the Hooded One*: "From Ymir's flesh the earth was created / And from his blood the Sea / Mountains from bone / Trees from hair / And from his skull the sky / And from his eyebrows the blithe gods made / Migdard, home of the sons of men / And from his brains / They sculpted the grim clouds."

The Maori, too, thought there had been a beginning. The father sky and earth mother were inseparable lovers locked in an embrace, but their passion was the curse of their children, trapped in darkness between them. After a time, the sons would have no more. Rongo, the god of cultivated food, tried to push his parents apart but could not break their love clinch. Tangaroa, the god of the sea, and his brother Haumia-tiketike, god of wild food, joined in, but they too failed to make a separation. Only after many attempts, lying on his back on his mother and pushing with his powerful legs against his father, did the god of forests and birds, Tāne, finally force his parents apart. For the first time, light and space came to the world and the sons were happy. But one son, the god of storms and wind, Tāwhirimātea, could not bear the sound of his parents' cries and vowed to avenge their sorrow. Ever since, hurricanes and thunderstorms and rain and mist and fog and whirlwinds have troubled the earth and its seas and fields and forests and fish and lizards and humans. And the sundered father sky, Rangi, and mother earth, Papa, forever continue to yearn.

This is not all. The Chinese believed the world hatched from an egg, and Aristotle that it was eternal.

3

The Universe was born approximately 13.799 billion years ago. It was not from a chicken or a dismembered corpse or a broken love embrace, but from a Big Bang that the Universe came. Nor was it just our own heavens and Earth that were created: our Solar System belongs in a galaxy of billions of stars, itself a galaxy among billions of galaxies. On large scales, the galaxies are uniform: there is no edge, nor heart, to the Universe.

4

First there was nothing: no Time, no Space, no Cause. Lucretius said that nothing can come from nothing, but he was wrong, according to the Scientists. For the Big Bang came from nothing at all, heralding the arrival of the forces that would seed and carve the Universe. One day they would be named the Weak Force and the Strong Force and the force of the fields, Magnetic and Electric, and the scrawniest force of all, Gravity.

When Time began after the Big Bang, well before the Scientists, all the forces were one, none yet stronger than the other. The forces were still united, but unity would prove fleeting. It was the Planck Epoch, dense and sweltering and dark and symmetrical. Epochs are long; this one lasted 10^{-43} seconds. After that the forces demanded to separate.

Gravity, suddenly the weakest of all, was the first to break away. What did she have to offer? After all, the Universe was only 10^{-35} meters long. But then the Strong Force stripped away, too, and there came the great Cosmic Inflation. In an instant, the Universe was as large as a grapefruit. Gravity would one day have her comeuppance: in this expanding Universe, true power would be exercised at a distance.

Great wars followed. In the cauldron, Matter and Anti-Matter became adversaries. And so when the Gluons gave rise to the Quarks, the Anti-Quarks marched out to fight them. Only one in a billion Quarks survived the onslaught, a narrow escape. From this remnant all matter would form.

It was time to consolidate. The Higgs Boson had already made mass possible. Stretched to a billion kilometers in diameter, the Universe had now cooled down to a mere trillion degrees. And so as Gravity looked on from afar, powerless, the surviving Quarks summoned the Strong Force to pull them together, giving birth to the Hadrons. No sooner had the Hadrons come into being than Anti-Hadrons materialized to annihilate them. Once again, just barely, Matter survived.

From a speck of nothing, the Universe had grown 100 billion kilometers in caliber. Looking at the clock ever since the Big Bang, incredibly, Time registered one second passed.

5

The path was now determined, just an instant from the start. And as the Universe continued to cool, the Hadrons stayed close together, fusing to form the first stable elements: Hydrogen, Helium, and trace amounts of Lithium. After twenty minutes, abruptly, as if scorned, they shut down all nuclear fusion, too cold to be able to stay together. For 380,000 years the Universe would drift in darkness, all wavelengths of light immediately absorbed by the free Electrons, a dense, searing plasma of formlessness. Gradually, as the freed Electrons were captured by the Atoms, Photons would decouple from Matter, escaping, traveling long distances before scattering, rendering the Universe transparent for the first time. Gas clouds would condense. Over billions of years, stars would form, and galaxies. The Dark Age would end. Heavy elements would be born. Light and Life would come to the Universe. And from them, eventually, Love.

The path was determined, but there was a catch: against all odds, the Universe continued expanding. It was Hubble who saw this, in 1929. Gravity had pulled the cosmic constellations into place, but the galaxies were spinning away from one another, the aggregates of all our future joys and sorrows disbanding.

Measuring the rate at which the Universe was expanding, two generations after Hubble, the Astronomers in 1998 would discover something even more astonishing: the expansion was not slowing down but accelerating. It was as if an apple had been hurled into the sky and kept on going faster as it rose. If they could, the other forces might have chuckled at the spectacle: Gravity had chosen distance as her ally, but distance had betrayed her, laughing in her face.

From beyond the grave, Einstein came to the rescue. Gravity, his mathematics showed, can push things apart as well as pull them together. Great truths are one and the opposite, just as Bohr had thought. Einstein didn't believe it at the time because no one had seen it. But the Universe had created a mist that made expansion possible. And soon we would give it a name.

6

Dark Energy was beguiling.

After all, to generate the repulsive Gravity for the accelerated expansion of the Universe, Dark Energy would need to exist in precisely the density of Planck units revealed to us by the Astronomers:

0.000000000000000000000000000000
00000000000000000000000000000000
00000000000000000000000000000000
0000000000000000000000000000136

Even the smallest deviation in this cosmic number would forever change reality: subtract one zero after the decimal point, and the Universe would be so dense that the galaxies would collapse in on themselves; change the 6 on the tail to a 7, and Gravity would paradoxically be pushing out so fast against the mist that galaxies wouldn't form at all. The 300 sextillion known stars and 100 billion known galaxies are but a fraction of the Universe; with space expanding faster than light can traverse it, what lies beyond remains opaque. Still, one thing the Astronomers do know: with even the slightest tweak to the amount of Dark Energy, "the All," as well as the possibility for Love and every departure from it, would vanish like a morning mist.

7

It seemed rather capricious if not altogether irresponsible. Why should such weight be placed on Dark Energy's narrow shoulders? Many Philosophers and Theologians claimed they had an answer.

It was then that the Universe revealed its precious secret to one group of believers, or so these people thought. These men and women believed in Strings so small even pygmy fleas couldn't hear their vibration. Neither depth nor height nor width could suffice to carry their intricate melodies, but the music they played was the most perfect of symphonies, uniting all the forces.

For despite appearances, tucked away undetected in extra dimensions, it was the Strings who had pulled away Gravity and the Strong Force to ignite the Cosmic Inflation. It was they who summoned the Quarks and the Anti-Quarks, and after them the Hadrons and Anti-Hadrons. And when the Electrons were captured, and the elements were wrought, and the gas clouds condensed and the stars and galaxies were created, it was the Strings who invited Dark Energy, just so, no more or less, to reveal the Janus face of Gravity. No one had ever seen them, it is true, but the mathematics demanded them, and the implications were dramatic. For from thence all the loves and all laughter, all songs and all sorrows, all grudges and all grievances would come flowing.

8

But here was the lie, of cosmic proportions:
The Universe was not alone.

If the Dark Energy was to exist, at least 10^{500} other universes were necessary, each with Strings tucked in hidden dimensions plucking a distinctive Dark Energy melody of their own. There would be the universe with

0.000000000000000000000000000000
00000000000000000000000000000000
00000000000000000000000000000000
0000000136

density of Dark Energy, and the one with

0.000000000000000000000000000000
00000000000000000000000000000000
000000000000000000136

and another with

0.11111111111111111111111111111
11111111111111

and a fourth

56.253748372625183636302639 2022

and a fifth

3.14

ad infinitum. Each would have different dimensions and different music, different elements and forces, some say a different mathematics or even philosophy. Our Universe, by pen-and-pencil fiat, was just one in a vast crowd. It had not come from a chicken or a dismembered corpse or a broken love embrace, but from a Big Bang burped by nothingness. Thus every drop that had ever evaporated, every future heart that would be broken or healed—even logic and all departures from it—had been tricked into believing in their uniqueness.

Nor was this the end of it. For the nothingness that had produced the Big Bang was insatiable, a fuel that could not be extinguished. This is the gospel the String Theorists proclaimed: the Big Bang had birthed our Universe, floating on a Bubble, but there had been many Big Bangs, and many Bubbles, all birthed by nothingness. Nor could the Bubbles touch or ever know each other, some said—all 10^{500} of them with their 10^{500} universes.

The Dark Energy had let out an even darker truth. Without all its zeros after the decimal point, without its precise tail and our incredulity over its narrow shoulders, we would have never known about the Strings or Bubbles or Universes in the first place, nor could we imagine limits to our supposed boundless imaginations. But the dimensions of our own Universe that would bring about all future yearnings, the unseen vibrations that would mend and break again all hearts, the masses of the particles, the strength of the Strong Forces, the restored pride of Gravity—none, in the end, could ever have been intended.

Instead they were just one of infinite possibilities,
necessary to no one but us.

HUBRIS

THE CREATION OF THE SOLAR SYSTEM

1

I am worshipped across the lands. The Egyptians call me Re and believe that my tears made the humans. To the Greeks I am Apollo, born on the island of Delos and professing all futures from Delphi. At their temples in Tenochtitlán, on the other side of the world, the Aztecs make human sacrifices to me, Tonatiuh, so that I have the strength to keep holding up the Universe. In Rome I heal, and in Sumeria I am Utu-Shamash, sitting on a throne, beholding the land, the merciful God of Justice.

In different cultures I take on different guises: Once I am a falcon with a serpent wrapped around my neck, once a blue hummingbird. I am a dung beetle, a lion, a ram-head in the Underworld.

My brother misbehaved and so I hid from him in a cave, closing the entrance behind me with a giant rock. Then the evil spirits came out, and the world went dark. Unhappy, the gods decided to lure me back, throwing a party outside the cave and placing a mirror in front of it—this is what the Shintoists say. Accompanied by music, the goddess of laughter began to dance, and hearing the music outside, I grew curious. When I peeped through a crevasse in the giant rock, I caught a glimpse of my reflection in the mirror and fell in love. That is when I, Amaterasu, came out of the cave and light returned to the world.

This is what others believe. But here is what I know:

I am the center of the Universe. Around me the planets dance. Those who come too close burn at my fire; those too far suffer cold for their distance. But even there, none escapes me. Even in the cold far reaches, all are under my spell.

The planets circle me, in step, counterclockwise. Yearning for my warmth, they follow my every turn. The planets circle me, but I burn only for myself.

Everything revolves around me.
Nothing eludes me.
Light comes from me to the world.

2

That is what the Sun would say if it could talk, but the Sun, like its worshippers, does not know itself.

Here is the truth:

The Sun is a middle-aged, G-type main-sequence star. It was born 4.6 billion years ago, and will die 5 billion years hence.

3

Self-delusion is a drug.

In the beginning there was just a giant lazy cloud of gas and dust. This was 9 billion years after the birth of the Universe. But then shock waves arrived, perhaps from a nearby supernova, and the cloud collapsed into itself and began to spin in obedience to Gravity. And the more the cloud spun, the hotter and denser its core became. And as matter fell toward the growing core, following once more the dictates of Gravity, a blazing ball began to form, trapping all radiation, like a wild animal dragging food into its den. After 50 million years, a blink in the life of the Universe, as the matter condensed, the temperature in the core had risen to 10 million degrees Kelvin. Squeezed together as if in a vise, the Hydrogen in the ball began to fuse into Helium. Then came a giant explosion. This is how the Sun was born.

And so it was all Gravity's work, Gravity and its sidekick, Angular Momentum. And as the explosion rippled through Space, the planets formed in the Solar Nebula. First they were just refuse dust grains, blown-out ejecta orbiting the central proto-Sun. Gradually they gathered into clusters, then collided to form larger bodies, all spinning around the Sun like drunken sentries. In the beginning there were hundreds, even thousands of them, but only eight abided, with an asteroid belt running between them. The origin was unremarkable: as beer is made from barley, hops, yeast, and water, so the Solar System was made from gas and dust and Gravity and Angular Momentum. All the planets revolve around the Sun, but the Sun itself was never a design, just a consequence.

4

Arrogance is blinding, and yet this much is undeniable: the Sun is the center of the Solar System. Even beyond Neptune, past the farthest reaches of the Kuiper Belt, 7.5 billion kilometers, or 50 AU, distant—even there the pull of the Sun is felt. Perhaps it should not surprise us: the Sun's mass is nearly 99 percent of the Solar System's, and Earth would fit into it a million times. But that is not the end of it. For one light-year is more than a thousand times that distance. And objects two light-years away are pulled by Gravity to the Sun.

Yes, the Solar System revolves around the Central One—even that elliptical eccentric, Halley's Comet. Blazing between Mercury and Venus or out near Pluto on its journey, it too, like everyone else, is on the leash of the Sun.

Such facts have led to a fantasy of solar proportions: after all, Halley's Comet, like everyone else, is nothing but a jailbird. The Sun's pride is based on an unfortunate misunderstanding: It is not for love that the planets circle it. It is because they are hopelessly trapped.

5

If the Sun could speak, it would get things all wrong.

In truth, the Outer Planets are not cold due to their distance; it's the opposite—because they are colder, they are distant. After all, when the spinning ball exploded, it was the lighter gases with low condensation points that were blown farther away, whereas the heavier materials with high condensation points stayed nearer. That is why Mercury and Venus and Earth and Mars are made of rock, and are comparably tiny: there wasn't much metallic material in the Solar Nebula from which to form the Inner Planets. It is also the reason why the two closest Jovian Planets are gaseous, and the two farthest icy: for only beyond the frost line could the volatile ices remain solid. Jupiter and Saturn grew large, capturing enormous atmospheres, whereas Uranus and Neptune just froze.

No, the planets do not circle the Sun out of love; they circle from obligation: Saturn's 62 satellites, Jupiter's 69, Neptune's 14, Uranus's 27, are merely victims of the force of circumstance. So too are the satellites' satellites, and the satellites of the satellites' satellites, all the way beyond the Kuiper Belt. Wrought by Gravity and Angular Momentum, the Solar System knows only force.

6

And so, despite its claim to the contrary, the Sun is not at all special. An average star, just one of billions, it is wasting itself away on an outer arm of the Milky Way. The Sun pulls on objects as far as 2 light-years away but is itself 26,000 light-years away from the center of just one galaxy.

The Sun is brighter than ever, but don't be tricked. For every second that passes, every precious breath that you take, at its core the Sun is fusing 620 metric tons of Hydrogen into Helium. It would say that it burns for itself, but in reality it is eating itself. The Solar System's gleam is due to self-cannibalism.

And in 5 billion years, when the Hydrogen is depleted, despite the Greeks and Romans and Aztecs and Japanese; despite the human sacrifice and the underworld, the dung beetle and the blue hummingbird; despite all these, the Sun will die. First its core will collapse, in obedience to Gravity. Then its outer layer will expand, 260-fold, uncoupling from the core and turning into a Red Giant. And as Mercury and Venus evaporate and all remnants of life on Earth are wiped out, the core will grow so hot that the Helium will fuse in an instant. And as its outer layers burst into space, returning the elements that created it to an indifferent Universe, a tiny White Dwarf will remain, no larger than Earth, a shell and shadow of its former self.

Yes, the Sun's beauty is in full bloom. But it is a cut flower, despite it all.

7

The Sun fell in love with itself,
but like its many worshippers,
the Sun did not know itself.

We know all this
because we are the Scientists.

Light comes from us to the world.

MOTHERHOOD

THE EARTH
AND THE MOON

1

In the beginning you were so close, just 22,400 kilometers away. What days those were! Just four hours long.

I admit, when you first arrived I was shaken to the core. My body was one big wound, oozing a bloody magma, shooting up volcanoes, my every wave a tsunami. I wasn't sure I would make it, to tell you the truth. The shock of your arrival nearly killed me.

Men thought that I had snagged you, or simply spat you out, but they were wrong. Your birth was neither an act of abduction nor an expulsion, but a terrible tearing-away.

2

If you don't believe me, consider this: Had I snagged you, just like that, in midflight on your trajectory, I would have needed a gigantic atmosphere to slow you down. But I had no gigantic atmosphere for the purpose of capture, nor for any other purpose. It was by a different course that you arrived, forever changing my world.

Consider this too: Had I spat you out, I would have had to be spinning so fast that there would only be day, but this was never the case. My days were once much shorter, but, however brief, they were always followed by night.

And so it was a giant rock, not a purge or a furtive grab, that hastened your delivery. Ramming violently into my side at an angle, just so, like a bully, taking me by surprise. It tore away my flesh. The pain was unbearable. But I was overjoyed, despite it all. Despite the agony and my beaten body and the searing ocean red with magma, I was elated. For there you were, suddenly: Of me, but not me. Coming from me, but not only me. And I knew that from then on, even in my darkest hours, there would be light.

3

I speak to you candidly.

I was born from countless mineral grains in space that clumped together thanks to Gravity. Bishop Ussher dated my arrival precisely: at nightfall before Sunday, October 23, in the year 4004 before Christ. What a fool. And as if it were not enough that I was a chimera, forged around the whizzing proto-Sun, more than two centuries later

Lord Kelvin allowed between 20 million and 400 million years since my beginning. Courtesy of the New Zealander Rutherford and the discovery of radioactivity, more discerning men slowly came to fathom my ancient pedigree. Carrying a particular amount of lead isotope, a wedge of meteorite found in a desert crater finally divulged my secret, in 1953. I am more than 4.5 billion years old.

When I was born, I was devilishly hot. Pretty soon I melted completely. No water. No continents. Just a molten sea, 3,600 searing degrees Fahrenheit, and the heavy iron that quickly sank to my core. At my edge, space registered a biting minus 450; it didn't take long, therefore, for a thin crust to consolidate, disguising my internal inferno.

Slowly things quieted, as much as they could. And just as I was gaining my balance, that's when Theia, as large as Mars, slammed into me, unannounced, bringing about the Great Tearing-Away.

At first I thought it was the end of me. But Gravity came to the rescue, and once more that sidekick, Angular Momentum. Of Theia nothing remained, but you and I survived.

And when I saw you for the first time, circling me, I felt as if I had never lived before, as though I too had just been born, taking my first breaths. A cool sweat enveloped me.

You were round and glowing, filling the entire firmament. Neither Neptune nor Venus had such a gift to gaze upon, and Mars could hardly score those two scrawny prunes true moons. I was the sole terrestrial planet with a Moon of my own, the largest of its kind in the Sun's ambit.

4

Years passed by me. The iron had long settled to give me my core, and above it the mantle and above it, once more, the crust. But I was still a parched and rankled planet, devoid of Oxygen, wallowing beneath a poisonous atmosphere. For eons it remained a mystery how my wounds became comforted. Then a meteorite landed in 1998 and was, after a time, examined under the microscope. There they were beneath the ocular, tiny crystals of salt, and within them, even smaller volumes of liquid. Just as with Theia, this deliverance, too, had been a deus ex machina. The rock was more than 4.5 billion years old, and could no longer keep its secret: all those years ago, incredibly, water had come to me from another world.

I was still spinning at a wild rate then, my days just ten hours long. Under bombardment by meteors, gradually my seas formed with the aid of water vapor from volcanic eruptions, drop by drop like assembled tears. Yes, my waters were brown and muddy at first, wafting piteously beneath a deep orange sky; the Sun was then a dim brick-red orb, squinting beyond my coat of toxic gases. But soon a curious thing happened: as the oceans grew, lapping against my forming continents, slowly they began to find a rhythm, and I began to see that the rhythm was due to you. It was you, so close to me then, who governed the ebb and flow of my existence. Twice a day, as you passed my equator, fifteen times closer to me than you are today, the oceans bulged, as if expanding to greet you. But when you rose, and when you set, the waters shrank away, like a spider ducking into its carapace, so emotional at your comings and goings that they needed to hide.

Remember: when Theia hit me, uninvited, it tilted me on my axis. Suddenly, one of my poles was closer to the Sun and the other farther away. And so it happened that just as you were born, so too were my moods created: languid in summer, dark in winter, carefree in spring, pensive in autumn.

I wonder sometimes: Was it just coincidence?

5

Once, like me, you too spun. But soon you were locked in place by the tidal rhythms, our grip on one another tightening—action at a distance. Always, you showed me a bright face, never your dark side.

Funny games you learned to play with me. When I come between you and the Sun, you disappear entirely, and when it is the other way around, it is you who hides the Sun. I know your tricks are illusions: the Sun's diameter is 400 times larger than yours, but the Sun is also 400 times more distant. That is why you seem as magisterial. Perhaps it is also why you have stayed closer to the Sun's ecliptic, rather than mine, hedging your bets, instinctive about your true master. Still, in my heart of hearts—damn the physics—you loom large because you are my own, exclusively.

6

I cannot stop time.

Due to my bulging tides, the ones expanding as if to greet you, you're moving away now, at the pace of the growth of fingernails. You seem determined, an inch and a half at a time, to gain your independence. And as you move away, yearly, I feel my center losing its balance, my internal axis inching imperceptibly toward chaos, my spin slowing down.

I comfort myself. As the Bard has written:

> It is the very error of the moon;
> She comes more nearer earth than she was wont,
> And makes men mad.

And so perhaps it is for humanity's sake that you drift, not a child's rebellion. But then again maybe it is your dark side, opaque to me.

I am too young to know, and too old to find out.

IMMORTALITY

LIFE COMES TO THE PLANET

1

Ribozyme was the original Hero,
unknowingly, well before any others.
The early humans omitted him from their stories,
but his downfall relates earthly wonders.

And so we return to ancient beginnings,
and with them ancient surprises.
For when life first arrived,
however it was contrived,
it took on some very strange guises.

This is how it happened,
or at least what some believe.
For despite their importance,
Beginnings erased themselves.

Thus they continue to deceive.

2

Life began in the sea.

It was after relentless asteroids showered down and boiled away the oceans; after steam condensed and returned to Earth in primordial monsoons; after still more asteroids vaporized the oceans, and still more monsoons arrived to fill the basins up again. It was after the ultraviolet radiation simmered, finally, at the end of the loathsome Hadean. No wonder 700 million years of incessant bombardment had been named for the god of the underworld. Still, gradually the impacts relented—an act of mercy attributable to no one. It was then that it happened, beneath the sulking dark blue waves.

The Earth was venting. In the deep waters, fields of tubular smokers jutted from the buried crust above the mantle, spewing black fumes into the night. The smell of Sulfur was everywhere; there was not a trace of Oxygen. So scalding was the heat, so skull-breaking the pressure, no living creature would have dreamt to call it home. A far cry it was from the shallow, calm pond that Darwin had imagined.

And yet this is where it happened, almost 4 billion years ago, as the little yellow submarine *Alvin* would one day discover. Here, in the cavernous pungent forest of jutting hydrothermal vents, the ingredients of life materialized. For despite appearances, compared to the carnage above, the depths were a sanctuary. As hot mineral-rich fluids bubbled upward through the rock-lined fissures, the elements carried out their uninspired machinations. It was far from a drama, in fact hardly a spectacle: carbon monoxide broken down to Carbon, hydrogen sulfide broken down to Sulfur. There was no divine spark, no cry from the heavens. Drip by dribble, snip by imperceptible hitch, the molecules of life assembled. First life was not a free-living cell but a labyrinth of minuscule minerals; it arrived not from a thunderbolt but from a porous rock. As the iron in the mix turned slowly into pyrite, a gradient transpired, on fool's gold. And like an unexpected smile, the energy for life was born.

That was all.

3

Life began not in the sea but in the clouds.

Even before the core and the mantle and the crust, before the oceans and the locking of the Moon to the tides, before the vents and before the labyrinths, a haze of dust encircled Earth. It was then that clouds of water vapor came together, and within them tiny droplets, like proto-cells.

One after the other, asteroids pelted down on the fledgling planet. And yet what looked like a curse was actually a blessing. For the barrage that scarred her with craters also flung into the sky the lifeless elements that had gradually accumulated: the metallic iron and the carbon dioxide, the liquid water, and above all, the potent driver of chemical reactions called Hydrogen. And as they catapulted like shrapnel through the rays of the energy-giving Sun—unceremoniously discarded with every earthly thud—life assembled, just like that, in midair. And rising higher and higher, it tumbled into the clouds, and found sanctuary within the droplets. It was there the stunned molecules woke, in their new homes, incredulous at having crossed the boundary, irreversibly.

Soon the barrage quieted. And Earth cooled. And the oceans rose. And great rains fell from the sky, bringing life to the planet.

4

Life began on Mars. There, on that last of terrestrial planets. Just ask *Curiosity*.

Like Earth, Mars was being bombarded, but there was a crucial difference. For Mars was tiny compared to its cousin, just half the diameter and a tenth the mass. When the asteroids hit, its refuse went flying, and there was no Gravity to hold it back. Having evolved on Mars, life could thus escape, easily. There would be no trouble reaching Earth on a meteor, the larger planet drawing it in, unsterilized, with its magic molecules intact.

The conditions on the small planet were providential for life, undoubtedly. For while Earth was covered by water entirely, there were mere lakes on Mars, perhaps a few small oceans. And so the things that first became animated would not be torn apart by gushing gales, their premature celebrations crushed. Instead, at the feet of active volcanoes, bubbling streams percolated down alluvial fans, massaging the incipient elements, gently breathing life into them, as they stitched together. Crawling from crater to crater, at just the right temperature, slowly decanting and distilling, the waters conspired with the rocks, as if they both knew. Then the timely rock hit and promise went flying, never to return.

Earth wasn't mature enough for life 4 billion years ago. Instead, life came from Mars.

5

Whether wrought beneath the sea, in the clouds, or far away on the small red planet, life would need to solve the greatest riddle yet. The Earth venting, the billows pelting, Mars inventing—let the Scientists quibble, it mattered little. Life's most daunting challenge still needed to be met.

To turn a corner, life would have to start marching: unlike a rock, it would be obliged to learn to replicate. But more than just replicate, it would have to resist entropy: distinct from a sunlit dewdrop, it would need to make its own energy. Life would have to learn not one, but two tricks simultaneously: without both Reproduction *and* Metabolism it would never cross the threshold.

The charge was clear, but there remained an unbridgeable chasm. For the only thing that could metabolize would be unemployed if there was nothing to replicate, and the only thing that knew how to replicate would lie prostrate without a workman to shepherd replication. It was the original chicken-and-egg problem, according to the Scientists. Life couldn't start without one or the other, but one of them had to commence.

Forget about Hercules, Achilles, or Actaeon: Ribozyme was the original Hero, eons before the rest. For by fastening his own nucleic acids to their complements, he'd be able to replicate with no help from another. Makee and maker, jack-of-all-trades, he was primed to be evolution's founder.

And so, 4 billion years ago, Ribozyme came to the rescue, under the waves or in a cloud or from a small stream on Mars beneath a volcano. It was clear that his powers were extraordinary, almost a sin: twisting around himself like a snake, he could kiss his ends to make a twin.

Biology had arrived to join Chemistry and Physics.
At long last, the march could begin.

6

The march could begin, but immediately it transpired: exact replication was not a godsend but a superpower to be blunted. After all, with no variations to choose from, how could evolution commence? An unchanging Biology could not adapt to a changing environment. If copying was faultless, the Hero would drive Life to a standstill. If it was no better than random, Life would just devolve.

It was then that the Old Guards, Chemistry and Physics, reared their heads, as if inspired. They determined that rather than copying himself like a master draftsman, Ribozyme would err like a nervous typist. Instead of perfection, what proved necessary was carelessness, a reversal that would fatefully abide.

And so what might have been a formula for pristine duplicates morphed in practice into a concoction for mock-offs, bastardized versions but survivors. The ordered march had been a fiction; kidnapped by the mutation rate, a frolicking ramble materialized instead. Thanks to the Old Guards, the drama of descent was unfolding, unplanned. The road would remain smarter than its traveler.

Evolution triumphed, but not without a victim: Ribozyme, humiliated, wasted away from one replication to the next. There was no way he could see ahead now, plot where he and life were going. In the end he was just a pawn, not a Hero. Rudely he'd been manhandled by happenstance.

7

Adam and Eve ate from the Tree of Knowledge, and were thrown out of the Garden of Eden to prevent them eating from the Tree of Life. But Ribozyme's vagaries were precisely the opposite: because he ate from the Tree of Life, his descendants would need to forsake the Tree of Knowledge. This was the bargain that was struck many years ago: to get anywhere at all, life would have to climb blind.

Posterity would determine it a recipe for immortality. For with poor copies and worse ones, there were fit and fitter, suddenly. Ribozyme would soon be thrown aside like a rind for a sturdier bearer of heredity, but the principle had been settled: when it arrived to replace him, DNA too would inherit the required blindness. Thus, varying unfailingly, life on the planet would go on cycling according to the serendipity of mutations and the fixed law of Gravity. On the back of its luckless founder, with thanks to natural selection, from simple beginnings endless forms most beautiful began to evolve.

In a twist as heartless as it was insightful, ignorance of the future had become the future's own price. This, wise men say, was the end of all meaning in the world.

And the wiser still—only the start.

LOVE

THE WEB OF LIFE

1

Consider *The Most Pleasant and Delectable Questions of Love*, written by Boccaccio the Florentine.

Two bachelors are arguing over which of them a young maiden loves. They turn to her mother—perhaps she might elicit an answer. Before long the virgin walks over to the nervous bachelors. Taking off a garland of leaves adorning her head, she places it on the bare head of one, unblinking; removing a garland from the head of the other, she slips it gently on her head without a smile. Then she turns around and takes her leave.

The judgment has been given, but whom does the maiden love? It is I, says the one, assuredly; why, unquestionably, announces the other, it's me she loves.

True, says the Florentine, the maiden has taken the garland from the second man because what is his pleases her. But truer still is that giving entails a receiver, affirming the beginnings of amity and love. Perhaps the maiden liked both men and was conniving to ensure the love of one without losing the love of the other. But as clear as Dido bursting into flames on her pyre, giving is the surest sign of affection. For why would the prima donna of Carthage shower gifts upon her heart's desire but demand none in return? Knowing the answer, she had lacked the courage to inquire.

2

The sign of Love is to give, not to be presented,
Such was the Florentine's resolution.
But this, don't forget, was the Fourteenth Century yet,
And no one, including the Florentine,
Knew a thing about evolution.

Here is the real story of how Love regressed
And also how it trucked and it traded.
For when it arrived
It was dumb but alive
But when it progressed
It just faded.

It is true, oftentimes
What seems obvious is misguided
And what's clear to us rather opaque.
For Love, in the end,
Has not a thing to portend
But that survival comes well
Before heartache.

3

Before the barnacles and bandicoots and peacocks and late Romantics came the Age of Great Diversity. Drifting on the planet, RNA viruses collided with DNA viruses, RNA-DNA amalgams with Ribozyme-protein melds, lipid proto-cells with protein-DNA concoctions. The Greeks would later speak of a fire-breathing monster from Lycia who roamed the banks of the ancient Indus, its tail a serpent, one head a bleating goat, the other a roaring lion. This was quaint, but with all due respect, the chemical chimeras were infinitely more imaginative. After all, life had yet to settle on just one plan.

Before DNA alone was chosen to safeguard inheritance, the world was a curiosity cabinet of the living and nonliving at the nanoscale.

And at no time before or after would Love be so pure.

4

Here is the truth: the deliberations of the Florentine are a farce.

When life began, Love was not a search for happiness but a strategy for survival. An imperative rather than a choice, the scheme proved farsighted. We call lovers those who share feelings, but back then, lovers were those who exchanged genetic materials. And when the many forms of life first drifted in the Age of Great Diversity, there were no scorned lovers, only lovers who had never met.

Blind to distinctions, the lovers traded wares dispassionately: DNA garlands for RNA sleeves, Ribozyme gloves for lipid sandals—an unthinking barter economy. It was a horizontal world, not a vertical one, fit for a future Portunid rather than a climbing Wisteria.

Love arrived well before the heart.

5

After a time, the Age of Great Diversity subsided. Thanks to the high mutation rate, the Antihero Ribozyme that had gotten evolution started was duly cast aside. Now DNA was primed to beat out all its rivals, its sturdy sugar-and-phosphate backbone reinforcing inheritance, the double helix winking at fate. Cells enveloped themselves in ever-discerning membranes. Order was emerging from disarray.

Before long, the lipid proto-cells were gone, and the DNA-RNA-Ribo-protein concoctions too had gone extinct. With permission from Chemistry, the zippered potentate seized control of heredity and, thanks to Physics, replicated with relative fidelity. Soon the upstart Biology was gaining a new confidence, increasingly resisting reduction to the Old Guard. With DNA at the helm now, it set about dividing labor: Ribozyme would be relegated to minion, lipids would become storehouses, proteins products. Gradually happenstance gave way to statistical regularity. And daughters, at long last, began resembling their moms.

The world order had changed, but one thing persisted: before the fuss of feelings, Love remained indifferent. For there was yet no recourse to what humans call sentiment. In the Age of Great Diversity, luck stood above all else: a trade that proved fortuitous survived, and a failed one perished—that was it.

To begin with, Love wasn't finicky; this was the original romantic spirit. Thus the Florentine had gotten it backward: true love is arbitrary.

Only later would come mere loves of choice.

6

Love's indifference wouldn't last long; the great tearing apart was imminent. Marching nowhere in particular, evolution began dividing life-forms into lineages. Genealogy, after chaos, had arrived.

The first act of partition was a dazzling division, setting the domain of Bacteria on the one side and the domain of Archaea on the other. Never before or after in natural history would there be such a schism. And soon it would forever change Love.

Following life's great rupture, the lineages were reluctant to separate, perhaps because their cell membranes and walls had grown more impermeable. Spurred on by the challenge, they developed a retractable tunnel. The Scientists would one day call it a pilus, and discover that Bacteria and Archaea bored holes into each other and rolled their DNA through them like an annular gift. Whether within Bacteria or Archaea or between them, rolling down the pili the ring-shaped DNA would become part of the other. Clawing on with its fingernails, Love had thus survived.

But there was a snag: these were far from indifferent exchanges. When the Scientists looked more carefully they found that they were shamelessly conditional. Within the spherical bestowal, encrypted, were the directions for building a pilus, and only once these directions were received could an Archaeon or a Bacterium seek out a mate. When they'd find one, duly, they'd pass on their donor's DNA complement. Here was the genetic arrangement: I'll make you a giver, and you'll perpetuate my lineage. It was quite a debasement, this new tit-for-tat world. For giving and allowing to give, precisely in opposition to the Florentine, were nothing but acts of self-love.

7

It was after Love had diminished from blind to reciprocal philanthropy that full-fledged Eukarya came along, a billion years after Bacteria and Archaea's arrival. It is true: in the beginning life's third domain would peer sideways, across the divides to the established lineages, for it had yet to gain the gumption to march on its own. But just before it brought about the Second Age of Diversity, leaving the older domains and its own doubts behind, just before the arrival of the barnacles and bandicoots and peacocks and late Romantics, Eukarya hid its DNA within a nucleus, rather than opening itself up freely, the way its progenitors had done. Eventually, when bodies were born, this would lead to the battle of the sexes. Alas, our ancestor had arrived distrustful. And so we, too, when the day came, would give and take our hearts selectively.

8

This has been the story of Love's denouement.

Many years ago it was determined that Life could not know where it was going, and so Life began to improvise. Improvisation brought serendipity, and serendipity—against all odds—the pure Love of survival. When the age of chimeras gave way to the age of DNA, chaos was replaced by genealogy, and two large domains began flourishing. But it was also the time when blind Love was replaced by mere requited exchanges—a contingent, thus degraded, form of its unfussy start.

It was later, when our ancestors arrived, that a further ignominy conspired to distance Love from its innocent beginnings. The Second Age of Diversity had come, and with it a third domain, Eukarya. Now DNA was suddenly hidden behind a nucleus. And as Life clambered upward rather than crawling to the sides, as it made bodies and minds, a tug-of-war began, fueled by suspicion. Fear and Lust and Envy and Shame and Vanity and Devotion were about to evolve.

In the dim days ahead, bewildered by the battle, men and women would invent drama, mistakenly believing, like the Florentine, that they'd found an insight into Love. If only they had paid more attention to evolution, they would have known that Love was infinitely purer before it became an emotion. Our intrigue masks a simple truth: the perpetuation of all that mattered used to hinge on promiscuity, and indifference.

Because of our hearts, despite our good intentions, the evolution of Life was the devolution of Love.

FREEDOM

SYMBIOSIS

1

It began with a timely invention. Not that anyone had foreseen it, or knew where it would lead.

These were the days of the Black Smokers, pungent hydrothermal vents in the depths of ancient oceans. The Sun's rays could not penetrate the submerged landscapes, nor could even Oxygen. Down there, there was only the smell of Sulfur. And a crushing pressure that felt like death.

2

Above, in the blessed shallows, a blue bacterium met a green pigment close to sea level one day; why this happened, no one knows. Immediately, Cyanobacterium and Chlorophyll fell into a dance so intermingled that the dyestuff became wedged into the microorganism's skin, like a crest.

Pouring down from the distant Sun, light rays stung Chlorophyll, embedded in Cyanobacterium. That's when it happened: the gaseous waters around them cleaved, energized just so, as if enacting a long-awaited deliverance. It was like the parting of the Red Sea, only more momentous, for this would be a universal salvation, not just a people's liberation. From carbon dioxide, Carbon now broke off to one side, while Oxygen, like an afterthought, fell to the other.

Later, in 1893, Charles Reid Barnes would call it Photosynthesis. And as the dancing pigment-bug basked beneath the rays of the future Re and Apollo, munching on sugary carbon hydrates, tiny balloons of Oxygen gradually rose into the atmosphere, slowly suffocating the noxious environment. There had been no scheme, not even a whisper, but pellet by pellet, the afterthought transformed the Earth.

3

At around this time the world was called to order: it would not do to hide one's head in a hydrothermal vent. And so Chance blew through the landscape, feared and uninvited. With neither target nor stratagem, a mysterious Oxygen-hating organism now rose slowly from the crushing depths. Some believe it was little Methanogen; others claim worse than mere ignorance: according to them, it will never be confessed.

If it was Methanogen, whether spherical or rod-shaped, it must have been astonished to happen upon the new breathing environment above: minerals had already drawn welcome breaths to form iron ores, and rich marbles began to color the rising continents. Until recently a bitter enemy, Oxygen might now morph into a godsend, if only Methanogen could learn how to make it its friend.

Aided by Chance and its more exacting running mate, Natural Selection, Methanogen now shed its cell wall and began to grow, remaining with a thin and pervious membrane. Frightened by its vulnerable predicament, tucking in as if to consolidate, a piece now folded and detached, inside. It was then that it corralled the DNA in its abdomen. We know what we are but rarely what we might become: from an archaic Prokaryote, Methanogen was morphing into a Eukaryote.

Still, if Evolution was on the march, in some grander plan unknown, it divulged nothing to innocent Methanogen. For minus a cell wall, and with a nucleus to safeguard its inheritance, Methanogen had no inkling of what would come next. If only it could have imagined the consequence of what lay ahead: Would it have recoiled? Would it have fled? The thought plays tricks on us. For if we answer in the affirmative, we would never have been here to ask the question in the first place.

4

Listen carefully, for this is what transpired.

Methanogen was wafting listlessly in the shallows of a forming continent. In the distance, a tufa column rose ominously above the waters. The sphere was quiet, serene almost. But the calm masked a deadly confluence. For with all the Oxygen now amassing, Methanogen was losing its breath.

Like a fish out of water, only in reverse, it suffocated on the Oxygen. The great rising from the hydrothermal vents had brought it thus far, so close, and the shedding of the cell wall, and the birth of the nucleus—all might prove to be a tease, a cruel and empty promise. Untutored and coarse, Chance could easily trick itself into believing that it had already achieved a sufficient modicum of success.

But Chance smiled on Methanogen that day. For it was then, against reason, that two brothers appeared, from the tribe of Alpha-Proteobacteria. Methanogen might have known: it was pounce or perish, hunt or suffocate—an opportunity god-sent. And so, with barely weighable ounces of strength, bleeding out its final drops of methane, it lunged in their direction, spent.

It was a prayer to a heathen Universe, an act of desperation. But lashing against the unsuspecting brothers, Methanogen managed to engulf and drag one of the Alphas beyond its membrane.

Alpha was trapped, but Methanogen was finally liberated. For Oxygen was Methanogen's enemy, but an ally to Alpha, providing its sustenance. And so instead of eating its kill, Methanogen became Alpha's taskmaster. Out of breath, it had swallowed an iron lung.

Being torn from its brother was a harsh payment for Alpha's bad luck, but serendipity would nonetheless prove enlightening. For as the years passed, the virtues of incarceration gradually materialized: what was lost in autonomy was gained in protection. Once the hunter and the hunted, Methanogen and Alpha now became codependent. And life could finally rise from the depths, really.

5

Eons passed and the engulfed Alpha morphed into Mitochondrion, with a 30-million-volt charge per meter across its membrane, equivalent to a bolt of lightning. Thanks to its efficient use of Oxygen, it became the powerhouse of the living world; as it breathed and provided energy and multiplied within Methanogen, its grateful unicellular host grew to become the *sine qua non* of all future animals. Morphing into an anemone, a fish, a lizard, and then a mammal, one day the former hunter became a boy who saw a girl walking gently through a hollow.

And as the boy hid behind a tree, his heart racing, aching with desire, he spotted a rose in the grass and smiled. Imagining its workings, he felt his grief evaporate. Duly, he plucked the flower and, summoning courage, walked over to the lass.

Nine months later as the lovers lay in a meadow, embracing beneath a tender morning sunlight, a louse climbed the leg of their cooing baby, once more by the machinations of Chance. That evening, they were surprised to find the baby crying, and they grew frantic in the middle of the night as he boiled and shivered, covered in sweat. They could not know when their spotted boy died in the cursed hours of the early morning that he was just the last of yet untold grievances. That more of Napoleon's men would die of Typhus than were killed by the Russian Army, that 3 million Irish would be cut down by it in the Great Famine, and the son of a President, and a girl named Anne Frank, yet full of dreams wasting away in a concentration camp, just weeks before her liberation.

This the grieving boy and girl could not have known as they wept inconsolably, cuddling their expired baby. But there was a shocking truth of which they were oblivious, an even greater ignorance that would be revealed by future Scientists. For the vicious parasite within the louse that had felled their loved one was not a stranger but rather kin to us, the yet autonomous descendant of the Alpha who retained his freedom. This is what happened when Methanogen pounced, praying to a heathen Universe all those years ago: Chance had separated two brothers, fashioning one lineage life givers, the other killers. The Alpha that had been trapped thought itself a victim but was mistaken. And now the brother who got away returned to exact his revenge.

6

Chance is rude, but its lessons can be tender: the gleam of liberty is a deception; the dishonor of incarceration, an emancipation. This is why we survive against all likelihood.

And why we continue to tremble.

DEATH

SEX

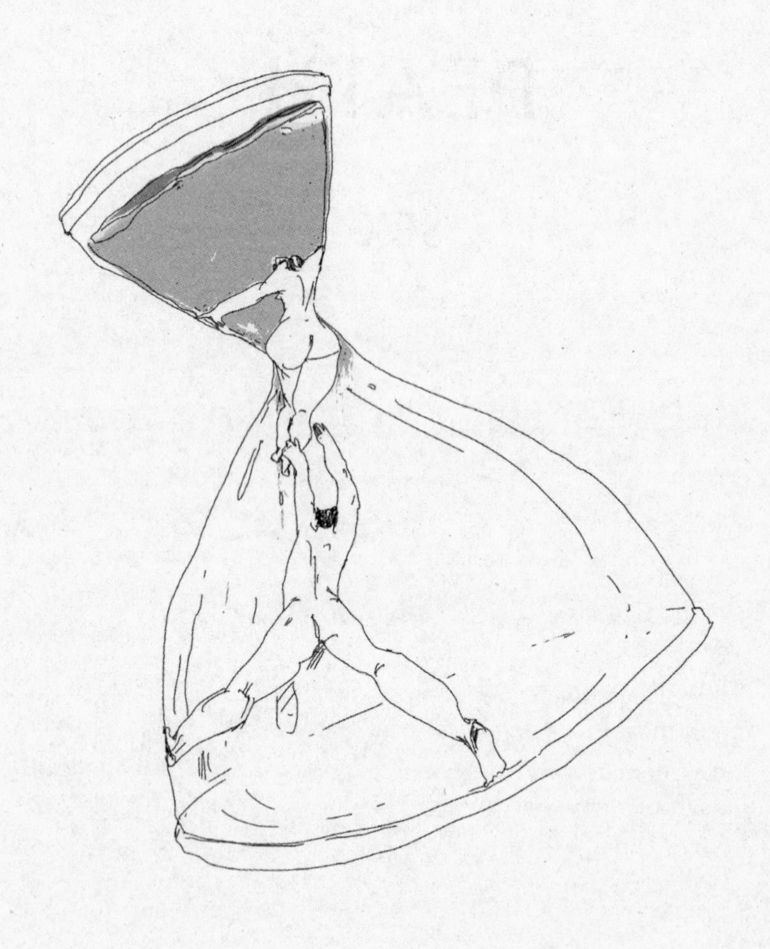

1

There was already night and day, and winter and summer; some say good and evil, though on this matter many dissent.

Why then, and why suddenly, the birth of the two sexes?

Why reduce the chances by half instead of a quarter or just one-millionth? It was difficult enough to find a mate on this Earth. And why surrender control over heredity in the first place? If life could beget life with no help from another, there might be no fathers, but daughters would smile much more like their mothers. Such a scheme would leave little room for bad judgment, and even less to wayward Chance.

2

Chlamydomonas wafted through the waters, oblivious. For what was there to fret over before gender had been invented? With no need to woo or make an impression, what worry was there in the world?

Her two flagella beat the current gently, her protean eyespot squinting into the light. She could not see much, but it did not matter. Like Narcissus, she needed no one but herself.

When the time came, she cleaved and split the difference. And there were two algae where there had been one, as simple as rain falling, or the arrival of dawn when the Earth turns to the Sun after night.

And yet there was a storm brewing, an aggregation of mishaps. For with every cleave, the DNA that was worn was passed on to the next generation unhindered. Soon the things that mattered most were being born flawed.

Lacking a brain, Chlamydomonas could not have known it, but this was a problem for which a solution was required. It was then that the tendrils of sex were unfurled—a flash of insight.

And one that would lead to our demise.

3

Mitochondrion stirred. With so much of Chlamydomonas's DNA already damaged, cleaving had become near impossible; healthy DNA being necessary for its workings, the wheels of reproduction would soon grind to a halt. If nothing was done, the fate of the tiny energy giver was sealed: a prisoner within Chlamydomonas, Mitochondrion would perish with the sinking boat.

Mitochondrion could sense trouble and panicked. Its host Chlamydomonas was a great-great-great-grandson of Methanogen times a thousand, and the original abduction had been a blessing, or so Mitochondrion thought. Had it all been a sham? A promise that fizzled? A nervous wreck, it let out a cry of toxic Free Radicals, hoping against hope at the very last minute to bolt.

And as the Free Radicals burst into the nucleus of Chlamydomonas, they wreaked havoc upon her DNA like a band of brigands. She could no longer cleave, and continued her lineage unaided: to survive, the one-eyed host would need to find another like herself and fuse, immediately. The price would be steep: her control over heredity would wane, and her splendid independence would be shattered. Yet it was plainly unavoidable. With two sets of chromosomes instead of one, there was still a chance that an undamaged copy of the genes might be passed on to the next generation. Mitochondrion, the instigator of the fusion, would in any case live to fight more battles, having escaped by compelling its erstwhile bachelor host to mate.

It is true what the ancients say, that there had been a world before this one. DNAs had rolled from kingdom to kingdom, giving birth to genealogy. But all this had been a kind of tease, like *coitus interruptus.* Despite reproduction, the world remained genderless.

Now all that was about to change.

4

Before the transformation, Chlamydomonas had brokered a peace. For years, with every cleave Mitochondrion had tripled within her, growing into a little army. And as its forces expanded, the DNA in Chlamydomonas's nucleus and the DNA in the Mitochondria negotiated the terms of symbiosis. Gradually, a balance was struck: Mitochondrion's army would provide the energy for Chlamydomonas's exploits, and in the verdant cytoplasm beyond its nucleus, *quid pro quo,* the host would provide its spacious barracks.

Except that presently, with her nuclear DNA damaged and Chlamydomonas unable to cleave, Mitochondrion tore up the contract. Casting aside years of coexistence, it began spewing out toxic Free Radicals—the army was going rogue. Alarmed, Chlamydomonas began seeking out an alga to save herself, and soon zeroed in on an unsuspecting target.

They were equal in every way, these hurried lovers that had not consented, the world's first bride and groom with no genders of which to speak. For Chlamydomonas's prospective mate had an army of trapped Mitochondria in its cytoplasm, leaking Free Radicals as it sought to jump ship. Like her, it too was being ravaged from within. Thus, alongside the DNA in their nuclei, both algae would bring with them full mitochondrial dowries. And as their membranes burst, their soldiers clenched their fists.

It was then that they resolved to do everything in their power to prevent the bloodshed, for a savage battle between their armies would surely mark both bride and groom's demise. The shame of it! How quickly lovers had turned rivals. What would it look like if the first-ever wedding became a massacre? The deed had to be done quietly. Thus both algae feigned pleasantries, but secretly sent molecular poisons to kill the soldiers in the other's phalanx.

In the end, the battle was won just barely. The rate at which Chlamydomonas silenced the other's Mitochondria minutely exceeded the reciprocal rate of her mate. Ninety-nine of a hundred had been felled within her, against one hundred in a hundred in her lover. But the one who survived saved the marriage.

For having silenced her mate's army while leaving untouched his nucleus, Chlamydomonas restored order; with the troops reined in and her DNA fortified, the lineage could be salvaged. The fusion between the nuclei marked the consummation of the marriage: double chromosomes meant double genes, and double genes meant survival. The world's first act of sex was an insurance policy for the ages. And life could go on marching undisturbed.

5

Sex set a precedent that would not go unnoticed: life's former parity would have to be abandoned. For to prevent battles, when two lovers unite only one can bring full dowry. And so future brides ballooned over time to countenance their growing mitochondrial armies, and future grooms increasingly shrank, replacing their armies with a motile tail instead. This is how the egg and the sperm were born, the moment that gender differentiated. It was also when equality was shed.

Later the organs of pleasure would evolve, and the world would come to know poetry, and fashion. But the origin of gender was artlessly prosaic: male and female, despite future extravagances, were created as damage control against the life throes of a nervous symbiont.

6

Years in the future when single cells came together to form communities, the game changed again, dramatically. For when life ceased to be a singleton affair, as it had been for

Chlamydomonas, egg and sperm could be sequestered to do the future's bidding, leaving behind the rest. Reproduction would depend on sex now, uniquely. A new order had forged an avant-garde.

Now Mitochondria could again concentrate on their host's survival: somatic cells that might burden reproduction would need to be eliminated. And so, purging the body of corruption by releasing Free Radicals, the symbiont became a sheriff, a metamorphosis of apoptotic proportions. Once again cells were betrayed by their tiny internal denizens, but this time with a collaborative, deadly twist. To reap the benefits of sex, the collective demanded individual sacrifice, and Mitochondria, which had brought it about, had the tool to impose it.

It was the German biologist August Weismann who would provide the abiding metaphor: Life is like a river. The banks, representing the body, are forever falling to the sides, never eternal, but the waters, representing the germ line, flow unbroken from future to past. Courtesy of Mitochondria, sex in a multicellular world had introduced a two-tier economy: like the flesh surrounding the seeds of fruit, bodies became shells, protecting the gametes. And when the seeds are planted the fruits can rot.

Mitochondrion ushered in a revolution: the energy that it brought with it catapulted life toward the heavens. Before long, the heat from its respiration would release organisms from the tyrannies of the seasons: warm-bloodedness would be born, and with it a night life. North and south zones would be colonized, and the planet gradually inhabited. For all the chaos Mitochondrion had created, from a single alga brains with 86 billion neurons developed. The symbiont was making amends.

7

Can reward exist without punishment? Nature was the first to shout out its reply.

For as Mitochondria accompany us through birth and adolescence and adulthood and procreation, the electron transport process that produces their energy begins slowly to break down. As with all machines, wear and tear has kicked in: loyal companions until now, Mitochondria will unwittingly start leaking Free Radicals. The results are inevitable, to some painfully grotesque: we will be nudged, against our will, into the indignity of senescence.

And as the years pass, we discover that our step is not as quick, that our eyes are losing focus, that our brains are becoming confused. The joys of the body are slipping away from us. Time's arrow will not be reversed.

8

It had been an unexpected journey for the micron-long evolutionary hitchhiker: first it was swallowed, then it made its host mate, then it became its keeper, then it led to its unraveling. It had played many roles—neurotic, pimp, sheriff, Thanatos—and brought the world sex, gender, complex life, and death.

Thanks to Chance, and Newton's Second Law of Thermodynamics, Nature's stratagems render philosophy and religion stale. Skeptics and believers alike may have known that despite the sins it often occasions, sex is Life's savior.

But they could not have known that its punishment is mortality.

PRIDE

THE ORIGINS OF MULTICELLULARITY

1

Hear me, Homo sapiens:

Summer came and went, and winter arrived, angry. A bitter frost covered the land, and I could feel the great oceans freeze. Flattening their bodies, multicellular beasts would one day burrow beside me into the silt, waiting for coming storms to pass. But there were yet no meter-long Dickinsonia, tiny spicule-fingered Fedomia, nor the ghost-like Spriggia, discovered by Reg Sprigg in the hills of southern Australia, serendipitously on his lunch break, in 1946. Such mysterious animals were still unimagined, shadows of my unmapped future and your forgotten past.

Not long ago, Rodinia had raised its head above the waves and shivered, as if doubting the wisdom of its own arrival. Furious winds were followed by stillness, and night crept over the horizon, fretful and glum. Mechanically, the cold seeped into the roots of the waters that would one day give their names to rivers: Meander, Melas, Ganges, Nile.

How could I know that there would be reason for hope notwithstanding? That after Ribozyme twisted and doubled, after Archaea pricked her pilus into Bacteria and let her DNA roll, after Alpha was swallowed and his brother got away—after all this, and still many other adventures, peace would come to Earth? You would name a future paradise, but look carefully at the fossils: eons before you arrived, it was the Garden of Ediacara that was the original Garden of Eden.

2

I am the father of Dictyostelium. A billion years ago, more than 300 million years before the Ediacaran, I slithered from the waters onto the escarpment. Slinking over a felsic crag, its quartz gleaming, I dived into Rodinia's warm turf. There were green algae in the currents I left behind me, and multicolored sheaths of the softest chiffon—the microbes—lying flamboyantly on the seafloor. But I would exchange my watery home for a dry birth. Before any life-form roamed there, well before the cold hit, I would be the first to colonize Earth.

In the beginning it was heaven. The territory was all my own, not a plant or animal on the landscape. Gliding through the moist soil, I was a conqueror, in want of nothing and no one. I could sense the Sun replacing the Moon at dawn, and feel the stars flicker in the sky at twilight. It is true, I was only a unicellular slime mold. But I was king of the continent.

Then came the cold, with its bitter winds to crush the spirit. And as the Sun was eclipsed, in almost an instant, the great promise turned into a curse, the land of possibilities into a betrayal. Like a black hole, the past now swallowed the future. Gone were the bacteria, my faithful diet. Numbed into oblivion were the tiny algae, my snacks. If beasts would one day roam these lands, as they would the waters, there was no one to whisper the news to me within Earth's barren, despairing tracks.

3

Listen carefully, humans:

It was 850 million years ago when the Cryogenian hit, freezing away much that had breathed on the planet. Teeming seas now turned into dead blocks of water, trapping the living with the dead.

Rodinia had become a graveyard, its jagged rocks marking the catastrophe like grotesque tombs. So chilling was the cold, it would have broken bones if they had existed. And so I burrowed more deeply into the soil, and on the brink of extinction, I lingered.

It was then that I sensed for the first time, astonished: the world is too harsh a place to go it alone. It was like a betrayal, this casting aside of independence; what promise I may have hoped for had beaten down my ego entirely. I had to admit now: self-sufficiency was valiant, but in the end it was laughable. Life straightens the will of the defiant.

And so as an act of resignation, I let out a groan, in the form of a cyclic adenosine monophosphate. If I had been touched by vanity, this final cry marked a hard-won education. There is both dignity and arrogance in autonomy. But there is greater comfort in staying alive.

4

Almost immediately, at the pace of diffusion, my brothers smelled my groan and followed the underground gradient until they found me. I was close to death: my body felt like a rowboat under a massive wave, the waters having crushed the rudder, the oarsman crouching by the scuppers, praying, eyes hiding behind a weather-burnt forearm.

Soon I discovered that all of my brothers were equally despondent, all ten thousand of them if not more. Where they had been when I climbed on the continent I really couldn't tell you; I had thought I was gloriously alone. Perhaps each figured itself a pioneer, until Earth spoke up to correct us. How foolish we must have seemed now to one another.

There seemed no end in sight to the crushing arctic. Inching up the gradient of the cyclic adenosine monophosphate, my brothers now swarmed me. We could not know at the time that our coming together in that dark soil would lead to a transformation. Instead, compelled by conditions to aggregate, we just huddled together against the bitter cold.

My brothers, too, had been a world of purpose in just one cell, autonomous and sovereign. Now, forcibly clumping, every one of us became a digit in a grander design, bowing its head and relinquishing self-government. And as a slug formed from our tiny bodies, each now a single cell in a becoming giant, we could feel the freedom ooze away from us, transferred to a higher command. Being the very stuff of the transfiguration, like dots in a pointillist painting, none of us could see the higher plan.

The turf was freezing, but no tarrying could now be countenanced: if we didn't find a way out of this predicament we'd all be doomed. It was then that the slug began moving, sensing its way out of the frozen inferno.

After it had crawled for some days, and kept going for more nights, it stopped, inches below the surface. And a stalk began to rise out of its head, like the periscope of a conniving submarine, until it pricked through the turf, testing the cool breeze above. Before I knew it, instinctively, I was clambering up the stalk, and I could sense that it was the dying bodies of my brethren, sacrificing for my deliverance, on which I was climbing. When I finally emerged aboveground, the light would have blinded me, if I'd had eyes, but I did not. Finally, the Sun had shone its rays on the Earth after all the wintry darkness. To a spore like me, it seemed a deliverance, or at the very least an opportunity. And so, collecting into a ball with my surviving brothers at the tip of the stalk, I waited for a good wind to blow me away from hunger. And when it came, I took to the air.

That is how I survived.

5

Later, when our ball of spores landed in a greener pasture, we all dispersed and I returned to my independent ways. Once more I was a world of purpose in just one cell, autonomous, but my brothers and I had learned the trick: when next hunger came, we pocketed our pride. This became our perennial cycle—forfeiture of ego alternating with satisfied solitude, altruism replacing self-regard.

From the heights of your intelligence you would see one day that I had not been the first to invent multicellularity; the lowly bacteria, my diet, had done it before me. Still, the example was there for all to see, a billion years ago, and would not go unheeded. From Volvox to Omphalotus to Acromyrmex to humans, singletons would come together in evolution to form communities, in at least twenty-eight separate lines. Some, like the cells of the plants and animals, stuck, throwing their lot together irrevocably; others disbanded, forever lost to time; still others alternated, like me, at nature's caprice. Whatever its fortunes, the surrender of selfhood was a glorious invention. For from one had come many, and from many came wholes, and from wholes came division of labor, and from division of labor came society, and from society came the minds that would one day unravel the mysteries of the Universe.

6

And so it came as quite a shock when your Scientists finally made the connection. For I did not know when I climbed on the stalk that there had been a fierce battle between my brethren that day—that some had cheated their way to deliverance by climbing the stalk instead of building it. Perhaps even I was one of them; I could not be certain. Our cycle was not as innocent as we thought.

After the cold passed, life returned and slowly began to flourish. The beasts of the Ediacara appeared, and my descendants degraded the bacteria that broke down their carcasses when they died. Soon, the nutrients that originated in the mysterious Fedomia, Dickensonia, and Spriggia were freed up in the soil and ushered in more-elaborate life-forms. And as time passed and the Tree of Life ascended, you came to learn that despite my lowly station, what is true for a slime mold is true for Homo sapiens: community is threatened by self-interest, benevolence forever wedded to indifference. Just like me, you live and die with the consequence.

In a multicellular world there can be no other way.

JEALOUSY

THE INVENTION OF THE EYE

1

My ancestors could not see, but I can see—the world and its many splendors. I don't know if I'm better off for it. In fact, if I could go back in time and reverse the invention, I would do it in a heartbeat. These eyes have brought me too much pain.

2

After Rodinia, Pannotia became the supercontinent. And before long, it, too, began to break up. In the south, Gondwana hovered over the Pole, cutting off ocean currents. Laurentia and Baltica and Siberia were drifting north, and the Earth was frigid. Gradually, as the great glaciers melted, sea levels rose, flooding large shelves in warm, shallow waters. With Oxygen levels climbing ever so slightly, a new epoch arrived, unlike any that had come before.

Darwin's teacher Sedgwick would call it the Cambrian, after the Latin for Wales, where the first fossils were dug.

You know: we lived through it in real time, the two of us. And from the Canadian Rockies to Chengjiang to Sirius Passet to Namibia, the full scale of the explosion would one day make itself known. The Garden of Ediacara was tame in comparison. It was during the Cambrian, beginning 542 million years ago, that life turned into war.

3

In the beginning, my mother, without words, instructed me: *Be careful. You are soft. There are many things out there that will want to hurt you.*

How could I know what to look for, or look out for? I had no eyes, just pores. The waters brought me all tidings: lurking dangers in the shallows, distant dreams in the depths. A rift ahead, a cold stream behind—all would seep through, equidistant from my heart, like muted whispers. This was my world, at the tip of a wavelet, the sum of my desires, the wellspring of my fears and hopes.

Sometimes the rocks surprised me: angular, jagged, balustrade-like, cold. They were my first teachers, introducing shape and form. But as my nasal bone became sculpted and my sense of smell grew, new feelings showed the world to me: fragrances of approaching basins, granular tidings of underwater scuffles, drafty portents of landmasses shifting.

This is how I learned of the continents' stirrings, a world coming apart at its seams.

For millions of years, our family lived near a calcite quarry. We imbibed our surroundings, and with every deposit of the transparent mineral into our bodies, shade arrived, gradually, to complement the other senses. In the beginning my progenitors could only see intimations of movement, light dancing obscurely. But then one day the minerals assembled in me just right, like raindrops coalescing on a leaf to quench someone's thirst.

Suddenly, the world appeared to squeeze itself into rods the shape of hexagons. With the calcite nestled in the gap between the tissues serving as a refractive lens, aperture materialized and resolution decoupled from hole size. And as the rods grew in number, the world sharpened and came into focus. Like the beasts of the Ediacaran, my ancestors had been loafers, passively taking in what life prescribed. In an instant, this was over: unsolicited, sight had arrived.

But these were no swollen, colorless eyes of a Humboldt squid, or dazed gelatinous eyes of an amberjack or marlin. They were not the monocled eyes of a lobster, or the head-shrinking eyes of a fly, or the beguiled eyes of a lemur in a tree. For there were no trees or lemurs and no flies in the world, no lobsters, or marlin or amberjack or squid. There was no vitreous gel, nor pupils, either.

But there was an optic nerve, and a retina and a cornea, and more than a thousand compound crystal lenses, made of calcite rods assembled as a camera. Each parsed a distinct azimuth, honeycomb-like, inviting in a slice of landscape. This is how the world first materialized, in all its pixelated glory.

4

I became a creature of fear but a creature of faith, too, almost instantly. For all around me I could now see Opabinia and hungry Hallucigenia, ferocious Anomalocaris, with jointed limbs and heaving gills and double tusks throwing menacing shadows on the seafloor. The sights were sinister, enough to cramp the stomach. But they invited, too, a hint of deliverance. For if I could see a danger, I could also hope to avoid it. Ever since, fear and faith have been companions, like grotesque parasitic twins.

I looked here and there, calibrating my distance to things, my eyes having assumed the office of house mathematician. If I darted to that rock, there at the foot of the sea lily, I'd be fine, with ample cover to protect me. I admit, the calcite had fashioned me a carapace besides my eyes, the better to protect me. I had a head shield to go along with it, this too is true. With happenstance and Natural Selection I had conquered my mother's caution. But my armor was no kind of match for the teeth of the bigger animals. And so I bolted.

In an instant, I was tucked beneath the stone, grateful, the world oblivious to my beat and breath.

It was then that I saw you, with all three thousand of my hexagonal eyes, hovering in the Sun's rays above me. You looked to me a Tamar, or maybe a young Cassandra. If ever beauty is splintered, it owes it to this first refraction: compounded, dancing between the solar javelins, the seer unseen. I know little of my own emotions, but I could feel something mystical. Vision ushered in panic, but close behind it was lust.

5

I could see you were like me: segmented, tripartite. Like me, you had whiskered antennae, and branched limbs, and a ribbed thorax, and a head shield, and a free cheek, and a carapace. You could roll up into a ball if you needed to, and prick your spines, and molt to shed your coat, like I could. But most of all, the crystals of your calcite were stacked together just like mine to promise perfect transparency, layer upon layer, like the handiwork of a natural mason. Humans would one day use limestone to build the pyramids at Gizeh, and the foundations of the Parthenon. But none would be quite as exquisite as our ancient Trilobite eyes.

We both knew: in a sightless world we would have continued to smell and feel our way around. Intricate caresses and olfactory trysts would have spread at the expense of yearning glances. Ears might even have flourished, one day spurring the most magical symphonies. The language of love would have been written in verses of vibration, sonnets of texture, perhaps even poems of perfumery. There would be no winks, or acts of looking away. Surely, we would know nothing of "blindness."

But chance had given us eyes, perfectly pure crystal ones. And reading the world's mosaic through living rock, our eyes became kings to us, relegating the other senses to henchmen. We needed our eyes in those muddy seafloors, where vicious animals disguised themselves as seaweed and worms burrowed below—where neighbors were betrayers, and nothing was as it seemed. When eyes came to the world, the world quickened: casualness was replaced by conspiracy, and suspicion. Eyes were a death blow to trust: it was see or be seen, eat or be eaten. This is how the war began.

Yet I forgot all this, tucked in my hiding place near the lily, as I glanced at you with my ancient crescentics. I forgot the lurking Opabinia and the hungry Hallucigenia, the ferocious Anomalocari with their jointed limbs and heaving gills and double tusks casting hostile shadows. I forgot the sneaking thieves and footpads. Forgot war and predation. In the dim lights shooting through my crystals, I saw you. And you were all that I desired.

6

Scientists would later claim that with eyes came locomotion, and with locomotion competition, and with competition specialization, and with specialization speciation, and with speciation diversity and the grand explosion of life. It was the eyes, they said, that abruptly replaced the Ediacaran with the Cambrian, introducing all the phyla to the planet in a big bang. Eyes were a blessing, accessories to celebrate. And so they were invented, again and again.

Still, it was I who perfected vision, all those years ago, though other creatures with simple eyes saw imperfectly before me. It was I who first tested the invention, through rock, for all that it could be. Whether the Scientists are right I do not know, for I am an extinct trilobite. But I can say that there was no Cambrian Explosion. And what's more: that eyes are a curse.

For from my hiding perch in the murky waters, I could see that day that your stony stare was planted elsewhere; that the sea lily, a colorful Echmatocrinus, was flittering gloriously in the sun-drenched waters of the shallows. You could not blink, which made this all the more painful. For viewing its tentacles, nine of them swaying before you so gently, you were hypnotized, and there was no way I could get you to turn your eyes on me. Silently watching your every move, I could sense the scores of horny male trilobites around you, their dirty eyes gleaming from within their crevices, rubbing their antennae furtively, ready to pounce.

Perhaps this is how Hades felt when he looked upon Persephone in the Sicilian meadow, spellbound by the violets. Perhaps he, too, could not help it, needing to have her to himself, exclusively. But he was a god, and while I dwelled in the ocean's underworld, I was just an arthropod. Hades could have had anyone he desired; it was you alone for whom I yearned.

In my defense, you broke a heart I did not know I had then, but what was I to do? Like everyone else, I followed the dictates of competition. You became my victim by no fault of your own but by my Darwinian imperative, and my vision. Then all fell silent as I violated you in the peaceful shallows above my rock.

We lived for 300 million years, the two of us, longer than the dinosaurs. On all accounts it had been a successful experiment. I know that to look through my eyes is to see the world as fragmented pieces of information: trilo-byte might have been a better appellation. But, please, at the very least, allow me to apologize: I learned the hard way, through the eyes, about lust and survival. What a shame they turned out to be one and the same. For in the end it left me no choice.

And so, from the realm of extinction, I implore the gods of chance retroactively: take away my eyes and never return them. To hell with all the bloody life-forms, and to hell with evolution: I would rather die a million deaths in the jaws of Anomalocaris than see through them ever again.

CURIOSITY

ONTO LAND

1

What would it feel like?

In that dry expanse where Oxygen is thirty times more plentiful, and sounds race through space four times slower than in these depths? What is it like, where light travels freely, as far as the eyes can see, not like here, where it soon fades, swallowed into oblivion like something to be ashamed of?

How would it be?

To have Gravity pull you down unrelentingly, to dream you can fly, if only for a moment, rather than glide forward or just tarry in place? What sort of zone has denseness not halved but made thinner by a factor of 800, rendering buoyancy an irrelevance?

How to survive?

Where temperature jumps in just twenty-four hours more than it fluctuates in an entire year, down here amid the columns? Where the Sun and the Moon determine everything, and can't ever be escaped?

What is the world like on land, above these waters?

2

It would not let go of the mind. A mystery beyond comprehension. How, from a fertilized egg, does life assemble, ready-made?

In the 1800s, the German scientist Karl Ernst von Baer first caught a glimpse of a solution. Examining chicken embryos one day, he made a surprising discovery. All the organs of the chicken, his microscope whispered, originated in one of three layers, migrating like birds until they found their rightful site in the becoming embryo. And what was true of the chicken was true of all vertebrates: all livers of all fish, all reptiles and birds and mammals, come from the endoderm, together with thyroids and tonsils and the lining of the gut. All heart walls emanate from the mesoderm, as do kidneys, and genital ducts. And all skins and nails and hairs and teeth enamel emerge from the ectoderm, where brains, too, originate. Thus every cry of a hawk, every mounting of a hedgehog, every heartbeat of a harlequin, every gaze of a human to the heavens—all came in development from the same place.

It was von Baer's countryman Ernst Haeckel who soon cried, "Ontogeny recapitulates phylogeny!"—a mouthful of a pronouncement and a wild notion besides. The handsome Haeckel thought he'd seen an undeniable progression: vertebrates, after conception, march through their evolutionary succession. Thus the human embryo begins in utero with gill slits, like a fish, then loses its reptile tail and finally its mammalian hairy coating. When it emerges, after nine long months, it is ready to be a human, having graduated in development through the history of its lineage.

The mind gnawed unsatisfied. Evolution might be true, but how do sperm and egg birth life in the first place? How from a dividing cell does the eye of the squid perfect itself, and the minuscule heart of the sparrow? In what kind of world, from a blob of undifferentiated cells in a uterus, will the libido of a hamster emanate, and the equations of Sir Isaac Newton?

Once again a countryman came *schnell* to the rescue: Hans Spemann could not sleep a wink until the riddle was solved. How, after all, does the embryo know at what end to build its head and at what end its anus? Where is the information stored? And is it physical? Or chemical?

Instead of starting with the mind, as Newton had, the German began with the egg of a newt. With a tiny hair from his baby daughter, he tied a knot around the fertilized egg, making two where one had been before. Lo and behold, perfect twin newts soon developed from the embryo. Manifestly, the information had already existed in the first to build a whole second animal. The question Spemann burned to answer was: Where?

The search continued unabated. Probing a salamander embryo in the lab one evening, just a sixteenth of an inch in diameter, Spemann's student Hilde Mangold cut off a piece of tissue smaller than the head of a pin. This she then grafted onto a separate developing embryo, and was overjoyed to find that it gave birth to itself and a conjoined twin. Mangold, not her mustachioed boss, had found the Organizer, impresario of development. Tragically, her surname proved fitting, for while she died from a kitchen stove fire before graduating, it was Herr Spemann who traveled to Stockholm to pick up the Nobel.

There were clues in abundance: legs growing out of the heads of flies, snakes with missing vertebrae, two thumbs or a tail born on unsuspecting humans. There was the Organizer of a frog, transplanted, that gave rise to a newt. Clearly, something in the Organizer was directing development. But what? The mind burned. And when the Revolution came, with Crick and Watson, the Scientists were afoot.

Soon Mangold's Organizer began divulging its secrets. That the Holy Grail is DNA. That the genes sculpt Nature's eggs, not hidden levers nor the grace of heaven. That genes mutate and can be turned on and off. That as creatures have an axis from head to tail, so the genes are arranged on the chromosome from top to bottom, internal cause mirroring external effect, incredibly.

But the greatest secret of all left even the Scientists mesmerized, for they soon learned that versions of the very same genes built all the creatures on Planet Earth. Up-Down, Left-Right, Ventral-Dorsal, head-to-anus: the center lines were laid, creating a road map for developing embryos. Thus, like conductors, instructions in the Organizer signaled cells to their fates irretrievably, in frogs and chickens and lions and mice. Even the sea anemone used the very same missives, creating symmetry deep beneath its unforthcoming, pulpy flesh.

The instructions were ancient, but they could be interchanged, like pieces of Lego: when a gene called "Noggin" from a sea anemone was injected into a normal frog embryo, two craniums of a frog appeared, like the mythic Amphisbaena. When the frog embryo's own Noggin gene was removed, waist and legs arrived punctually, like Sir Gawain's Green Knight, without a head.

These were the *Hox* ("homeobox") genes, top executives of development, named in deference to the Englishman William Bateson, who coined the words "homeosis" and (in 1905) "genetics." From sea slugs to fruit flies to humans, hundreds of millions of years ago they set the body plan in motion.

And magnificently, evolution would tweak the rest.

3

First came the jawless fish, wading for prey in underwater muck; they were the earliest vertebrates. Then the seas were filled like never before, or since: sharks sporting teeth-covered fins, armored placoderms, thirty-foot arthrodires, lungfish the size of elephants. This was the Devonian, and it would take with it much that it had invented. But the sharks and ray-finned fishes and lungfish survived.

The ray-fins were like sailboats, skin stretched between horny spines, carving through the waters. Powered by gills, they would grow to become 99 percent of all fish, from sturgeon to marlins. But the lobe-finned fish, among them the lungfish, had hard bones under their fleshy fins. All philosophy is destined to be reduced to bones, inevitably. But within these bones is also where it begins.

For there was an archaic common ancestor. And the ancestor had *Hox* genes. And the *Hox* genes laid an axis for the skeletal bones to form. And beyond the branches of the skeletal bones a ridge produced a small fin. And in this small creature's descendants, along the branch that led to the ray-fins, the *Hox* genes switched off earlier, while the genes at the base of the ridge turned on sooner, making larger fins to sail across the oceans. In the branch of the lobe-fins, meanwhile, the *Hox* genes worked much longer, so the axis was lengthened and the skeletal bones grew. Thereafter, delayed in growth, the fins became mere fringes, and the lobe-fish gained the look of living fossils.

And so Haeckel was right: it was all timing that led to our lineage. But Haeckel was wrong, since there is no simple succession of which to speak. Progress is an illusion, an invention to stroke our egos. Instead of a wise grandfather clock chiming a fixed ascent toward deliverance, there are hundreds of mindless timers tweaking cascades of development toward no particular goal. Where one alarm clock is wound up, another goes off and is silenced: the action speeds up one moment and falls behind the next. Here is the song that inquiry gradually chanted: as the drama unfolds, novel forms become experiments and—with luck—new species on Earth.

It was timing that led to our lineage, plus the vagaries of ancient habitats. For however clever genes may be, they cannot stick without selection. Thus when lungfish happened into brackish inlets and the waters of estuary woodlands, mutations to *Hox* genes began slowly taking root. For the shallows were spiked by tubers, and matted with dense vegetation. To navigate them, portly fins proved surprisingly useful.

And so bones beneath the flesh began elongating, until a shoulder appeared, and an elbow of sorts, and before too long, a wrist. And as the head flattened, as if persuaded by the shallows, the large hyomandibular bone ungripped its straddle, and a neck soon materialized. Then the hyomandibular shrank, turning into a stapes that would one day build an ear, inviting its bearer to listen.

In the age of the great beasts of the seas, it was grow big or get armor or die swiftly, a grim predicament. Escaping the ruthless placoderms, the ancestor of all children and explorers decided therefore to take another route. As if planning its getaway, its morphing continued. Slowly the rib cage expanded, to make room for lungs that had once been its swim bladder. Nostrils drilled cavities down to its mouth as it lifted its head above water and drew first breaths. Its limb buds were graduating, incredibly, into fingers. And when no one was looking, the fish rested its weight on its wrists and prepared itself.

The swimmer, against all odds, was about to walk.

4

Where to find the first ambler? The fossil men dug tirelessly as the mystery continued to tug.

First came Eusthenopteron, 385 million years old, unearthed in Miguasha and looking like a pike. But there was a squat head and stout lobes for fins—contradictions that seemed incongruous. This was a fish, all right, not quite a walker. But something undoubtedly was afoot.

Next came Ichthyostega, from the northern slope of Celsius Berg, with seven digits on its hind legs and ribs overlapping like venetian blinds. There was a tail, too, and a flat head and shoulders. The creature lived 365 million years ago and swam the waters, clearly. But this tetrapod was no fish, already.

Finally, diggers came to Ellesmere Island and, before long, sent their finds back home to labs in Philadelphia and Chicago. There, within the rock, a fish with a 375-million-year-old wrist appeared. Respectfully, the local Inuit were asked to choose a name for it.

And they peered into each other's eyes across time's frozen threshold.

5

Tiktaalik:

Did you wonder then what it would feel like,
to leave these waters, finally?
To breathe air, exclusively,
and ambulate on Earth?

6

The Lakota Indians of the Americas say that there was another world before this one but that its people misbehaved, so the Creating Power destroyed it with a flood. Alone in the world, Kangi, the crow, pleaded with the Creating Power to make him a new perch on which to rest.

From his huge pipe bag, which contained all the animals and birds, the Creating Power, obliging, chose four plungers known for their ability to remain underwater, and sent each out in turn to retrieve a lump of mud from beneath the waves. First the loon dived, but it could not reach the bottom. Then the otter failed, despite its strong webbed feet. Next, with its large flat tail, the beaver paddled valiantly, but surfaced empty-handed. Finally, the Creating Power reached for the turtle, urging the strange-looking reptile to return with mud.

When, after a time, no sign of the turtle presented itself, the heavens were sure it had drowned. It was then, with a splash, that he broke the water's surface, his face caked in soil and his feet and claws oozing mud. Grinning, the Creating Power fashioned an island from the mud just large enough for the turtle and Kangi the crow. Then he shook two eagle wing feathers over it until the earth spread wide and wider, swallowing all the water. Saddened for the dry earth, the Creating Power let out a cry, and his tears became the oceans, streams, and lakes. Reaching into the bag, he then dispersed all the other animals among the lands. And then he made the women and the men.

7

This is what the Lakota Indians say, but we know it was no turtle that first came upon land. It was the fish inside all of us, with its unsuspecting *Hox* genes.

This we know twice, because of our, and its, curiosity.
For one day it would make the women and the men.

SOLITUDE

INTO THE AIR

1

In the beginning it was known as TMM 41450-3. Then things got more personal.

Yes, life had already taken to the heavens, long ago in the Carboniferous, eons before TMM 41450-3's discovery in the spring of 1971. Some claim it was from gills that wings had originally sprouted, others that they grew from limbs. Some believe wings were a novelty, arising in tiny nubs that bud during development. There is little to go by; the guessing game rules. But whether wings on the Insects came from breathing or walking, in the waters or on land; whether they had or did not have a prior history—they needed to lift but ounces. And this, in comparison to TMM 41450-3, was child's play.

It was only later that the true challenge came, in the line of the Vertebrates. For there was lift and drag and thrust to navigate, and the terrible weight of the bones. Yes, after TMM 41450-3 evolution would yet converge, copying the ingenious solution. But before the birds, before the bats, before the planes, it was the lizards who took to the skies.

2

Much had transpired since the first fish walked onto land. For after a time, by laying eggs, the tetrapods could forsake the waters completely. It was then that they split into two separate lineages: the sauropods that would lead to the dinosaurs, and the synapsids that would lead to man.

On the human side countless lineages rose and fell, their names exotic and now forgotten: Caseids, Ophiacodontids, Gorgonopsians, Venjukoviamorphs. But there was progression, too, along a signal mammal axis: nostrils that moved from the front to the back of the mouth, so that their bearers did not need to hold their breath while they chewed; legs that fit below the bodies rather than sprawling to the sides so that trotting replaced waddling, the up-and-down movement relieving the S-shaped slither, decoupling breathing from ambling. And there were the brains, principally, that swelled and grew longer in development, tearing out the loose rear bones of the jaw, fashioning the delicate middle ear, ballooning to create the neocortex. This, in the end, would make all the difference.

The other lineage began with small two-legged carnivores, but grew into the fiercest forms the planet had known yet. In the waters, swordfish-shaped ichthyosaurs cut through the oceans, together with the serpentine-necked plesiosaurs, some eighty feet long. On land the dinosaurs romped, curdling the blood of all the creatures around them. And 220 million years ago, lizard cousins of the dinosaurs began tilting their heads to the heavens. For soon the pterosaurs would take to the great big blue above.

3

John K. "Jack" Northrop was born in Newark, New Jersey. At twenty-one, in 1916, he became a draftsman for the Loughead Aircraft Manufacturing Company, founded four years earlier by the Loughead brothers, Allan and Malcolm. There was the "flying boat," setting the American nonstop record for longest seaplane flight, from San Francisco to San Diego. There was the revolutionary monocoque fuselage of "the poor man's airplane," whose folding wings allowed it to be stored in a garage. The company folded in 1921, but the Loughead brothers opened shop again in 1926. Now there were Wiley Post and Amelia Earhart, flying Northrop's Lockheed Vega monoplane into the record books. Northrop even lent his hand to the wing of T. Claude Ryan's M-1, a bird that would set in motion Charles Lindbergh and the *Spirit of St. Louis.*

Chief engineer now of his own corporation, Northrop seemed an endless fountain of invention. There was the A-17 attack plane for the Army Air Corps, and for the Navy, the BT-1 bomber. There was the world's first stressed-metal-skin commercial plane, the Alpha; the 200-miles-per-hour Beta;

and, to revolutionize "airmail," the Gamma. When the war came, it brought with it the P-61 Black Widow night fighter. With its radar-guided armaments it practically shut down Hitler's night raids, planting fear in the enemy and saving scores.

Northrop was obsessed with his airplanes: he was born to revolutionize aviation.

For Northrop's heart was in the skies.

4

But how to pull upward, into the heavens? It was the lizards who solved the puzzle, defying the dictator, Gravity.

First there were the bones to hollow, and their walls to strengthen with internal struts. Gradually, fortified, the walls could diminish until they were as thin as iron leaves. Then there were the hind limbs, which in an explosive spurt would launch their bearers upward: these would need to redouble and toughen. But most important of all was the lengthening of the ring finger, the fourth digit on the forelimb, once an equal among mates. For as it grew and extended, grotesquely elongating, a membrane of skin enveloped it and did not let go, forming a sail between itself, the shoulder, and the hind leg. Thus could the bones on each side become like a half-mast, supporting a wing surface, durable and taut. And as the wings became blade-like, growing aerodynamic, they thickened in the direction of travel and tapered behind. Spinning in vortices, air could travel more quickly now above the long narrow wing than below it, creating the pressure difference that would lift them into the skies.

Later, independently, birds and bats would copy the model, and, feeling clever, human aircraft designers would one day call it a cambered airfoil.

5

Early in the spring of 1971, a twenty-two-year-old graduate student named Douglas Lawson made a big discovery. Atop a sandstone ridge in Big Bend National Park in Texas, he uncovered the world's largest pterosaur yet.

Already in the 1800s they'd been unearthed from limestone, their names becoming legend: Dimorphodon, Nyctosaurus, Pterodactylus, Tapejara. Some had bulbous beaks, dodo-like; others, shovel-shaped ones crowded with teeth; still others, elegantly tapered ones as fine as sharpened quills. Some had giant crests, perhaps to woo their lovers, perhaps as heavenly rudders. There were those that snatched flying insects in midair; others, fish from the surface of the seas; still others, cockles in the estuaries and shallows. They were insectivores, carnivores, maybe even scavengers. But taking to the air, surprisingly, they would all escape terrestrial predators, develop new ways to hunt, and widen the range for mating. It was from above that the pterosaurs discerned the Earth, and all its living creatures. Alone in the otherworld, for 150 million years they were masters of the skies.

The known pterosaurs, more than one hundred species strong, ranged from large to the size of sparrows, and yet not one was as colossal as this lizard-bird mammoth of the Cretaceous. Unbelieving, Lawson went digging, and soon found what he looked for: the thin, drawn-out plow-shaped skull, unearthly, with the eyes set like a toucan's at the base of the crested head; the serpentine neck vertebrae, longer than could be imagined; the hollow bones and supporting struts; the fossilized fur, indicator of warm-bloodedness; the quadrupedal launch pads. And finally the fourth finger, elongated beyond comprehension, until it formed a strut that supported a massive wing, fifty feet across.

It was clear: on land, the giant creature stood upright on its forelegs with its wing membranes folded like umbrellas, as tall as a giraffe. When it flew, exploiting thermals, it could travel ten thousand miles in a single flight. More fantastic than any dragon, the tailless beast pushed the limits of biomechanics. It was the greatest flying machine ever invented.

His hand shaking, Lawson scribbled on the canister: TMM 41450-3.

6

Sure, there was the F-89 Scorpion all-weather interceptor, the first American nuclear-armed fighter. There was the X-4 Bantam research plane that Chuck Yeager used to investigate near-sonic flight. Later there would be the SM-62 SNARK, the country's first guided intercontinental ballistic missile; the T-38 Talon; and the F-5 Tiger. But despite all these, there was only a single success that mattered. For there was always one dream in John K. Northrop's soaring heart.

Tailless, no fuselage, high-lift, low-drag—early on, Northrop determined: he would build the world's first "Flying Wing." A plane without a tail? To most the thought was laughable. But Northrop knew that the heavens cried out for it. And he would not disappoint their gods.

It was during World War II, when the Earth was scorched and burning, that Northrop began to design the wings that would first take to the skies. One after the other they arrived, in succession: the first all-magnesium, all-welded aircraft, the XP-56 fighter, and after that, the XP-79, its pilot prone. Then came the four-engine-powered B-35, "poetry in motion." Without the weighty fuselage, the tail surfaces, and the bulky engines and cowlings, rather than overpowering drag the flying wings reduced it. Soon they could deliver the same payload 25 percent farther, faster, higher.

A sliver in the sky, they'd be difficult to shoot down or detect by radar.

But Northrop's greatest plane of all was yet to arrive; it would come when the din of bloodshed had subsided. Its climb rate and low noise and vibration levels let it glide almost silently, 40,000 feet above. With a range of 10,000 miles and a wingspan of 170 feet, it was a knife's edge cutting up the sky at 464 miles per hour. Climb, bank, dive: its dual hydraulic systems made operation easy. For the B-49 bomber was the most awesome bird ever to climb into the sky.

Northrop's vision, alas, would not in the end go into production. Mysteriously, the U.S. Air Force ordered the flying-wing bombers on the assembly line scrapped, the existing prototypes destroyed, the jigs and the dies abandoned. Instead, a more conventional airplane was duly contracted to the competition: the B-36, with a fuselage and a tail like all the rest of them. Northrop retired prematurely and never recovered. Despite all his triumphs, he went to his grave a broken man. For there was one dream to which he'd fastened his desires.

7

In the end there would be a double comeuppance, or at least a bittersweet ode to solitude. For in his old age, Northrop received a letter from NASA: the flying wing had been rediscovered. Following his death, the B-2 Spirit stealth bomber would be constructed, the fiercest aircraft the world had yet known.

A year before that letter came, an even greater gesture was occasioned when, according to the dictates of Linnaeus, TMM 41450-3 was given its proper Latin name. Lawson christened it *Quetzalcoatlus*, after the feathered serpent god of the Aztecs, born to a virgin who had swallowed an emerald, or so according to legend. Representing wind and learning, Quetzalcoatl was the boundary maker and transgressor of earth and sky, a creator of mankind. But bowing to history, and the intricacies of human sentiment, Lawson gave it a second name—*Quetzalcoatlus northropi*—for the pioneer who had learned its secrets.

And yet, like the B-49, *Quetzalcoatlus* would be mysteriously discarded. It and Northrop had arrived before they were summoned; this was their tragic common fate. Prematurity would be both creatures' undoing: no lonelier destiny has been known to date.

The pterosaurs would go extinct, eclipsed by the birds, the only surviving dinosaurs. After the pterosaurs, reptiles would need to settle for being alligators in muddied waters, hopelessly sluggish turtles, and snakes slithering on their bellies in the dirt. And while the birds would flourish, and after them the bats, never again would nature match the flying lizard's lonesome majesty.

This, perhaps, was by some cosmic design inevitable. For when all was said and done, there was nothing personal about it, in the least.

SACRIFICE

RETURN TO THE SEA

1

I gained the world when you finally enveloped me. When the skies became a canopy to experience through an aqueous filter, only one emergence at a time. I had not been meant for this plunging, nor did I foresee or invite it. But I gained the world all those years ago, in the Eocene, when to hold my breath became my way of life.

Others would marvel at my size and sleekness and intelligence, but none was more surprised than me, for I had once been a rather dim dog. Perhaps it was an ancient hippopotamus—so the Scientists claim—my memory fails me. You, in any case, would be the conduit of my metamorphosis.

I wonder sometimes: How could so soft a medium bring about such harsh rearrangements? Then your cold touch against my body reminds me: it is our surroundings that shape us, despite ourselves.

2

There was a time when I needed ground below me, and this, I think, made all the difference. For I could lash my growing snout into rivers, surprise a fish, then backtrack onto land, satiated. I could even dive, for more than an instant, but I always knew there was *terra firma* behind. And so my nature was forged by predictability: whatever came afterward, it was the soil that shaped my soul.

So I want you at least to know, in the interest of fairness: I had no choice but to leave land behind all those years ago, nor was I aware of it until it was too late. Your expanses were profound, your opportunities more plentiful, and so I followed, beyond the edge, innocent and circumspect. As I barked to the Moon, overjoyed by the catch, I could not have imagined that, like Jonah, I would be swallowed.

At first I felt alien, but as the years passed I grew more comfortable. Already my skin was thickening, the subcutaneous fat adhering, my forelimbs shrinking, my hind limbs pushing, first against you, then growing closer to my body, giving up the fight. My tail, too, had stiffened to serve as rudder, then finally graduated into a fluke. Massive vertebrae sprouted down my back, supporting my swim stroke. To keep my gonads cool, my warm blood rerouted; to keep the rest of me insulated, my blubber accrued. Up my skull my nostrils climbed, the angle between my rostrum and basicranial stem growing, my ears slowly detaching, surrounded by foam. Trunk elongating, forelimbs morphing, my figure growing more smooth: I was the son of Ambulocetus, son of Pakicetus. And, opening your arms to me, you were my muse.

3

They would say one day that the blue whale has the largest penis the world has ever known—eight feet long—that its tongue alone can weigh as much as what they call an elephant. They would say that sperm whales dive six thousand feet deep, hunting like stealth submarines for giant squid below. That gray whales migrate annually, on a map I could not have fathomed, from summer feasting grounds in the Arctic to calving lagoons in Mexico, a round-trip 12,400 miles long. That the longings of humpbacks are carried in song throughout your dominion, summoning lovers from far and wide to mate. That the sleek fin whale, the greyhound of the sea, cuts through the depths at twenty knots and more, unimpeded, and that the right whale will eat a ton of krill on a bad day. Lucian spoke of a whale 150 miles long in which was contained an entire nation. After him Donne would praise the Beast of Revelation: "His ribs are pillars, and his high arch'd roof / Of bark, that blunts best steel, is thunder-proof." This they would say of my future unborn relatives. This and many more astonishing tales.

But I would like to say that when my forelimbs graduated into flippers and my hind legs diminished into vestiges, when I had gained so much fat that I was buoyant, truly, for the very first time—when all this happened I was a fish, retrogressively, though nothing like the other ones, despite it all. I do concede: I had never been freer, never before more elegant. But there is a remembrance that lingers.

4

I don't mean to be ungrateful. I gained the world when you finally enveloped me, though I never meant to make of you my residence. This I did thanks to the speed of sound in aqueous mediums, four times faster than in the air above. And so, beneath your waters, squeezing the air in my nasal cavities, their passages running now uniquely above my eyes, I vibrated the muscles. And as a fatty melon grew on top of my upper jaw, the vibrations of the muscles passed through it, focusing. And the muscles pressed the melon into different shapes, purposely, the skull behind them rising to a ridge to prevent the sound from moving backward. Thus through my snout I blew out sound until I could see the world before me. For the sound waves bounced back, returning to hit my jaw, then traveled to my ear, loyally. And I began to make out the objects that comprised my universe: a coral bed, a school of fish, the seafloor looming. Scientists one day would claim my descendants could distinguish a triangle from a pentagon at the distance of two hundred yards, calculate speed vectors, even tell apart a disc of copper from aluminum. Once I could barely see through the open night, my eyes straining. Now my nose illuminated the darkest caverns of the underworld.

Ketos means "sea monster," but I was sea maker, infusing your province with speech. For the noises I sounded helped me see and navigate, but they helped me communicate, too, with members of my race. When your waters chilled, 35 million years ago, those who survived were even more discerning, with smaller teeth, perhaps, but bigger encephalons. And so, gradually, as my body morphed, so did my intelligence, for the dialects that your waters carried brought us together—to mate, raise offspring, and hunt. I had once been a rather stupid form of ungulate. But traveling beneath your tints of cobalt, sapphire, and indigo, majestic, I was now the smartest creature in the seas.

5

And yet, despite your welcome arms, I wonder: Did you notice?

That each time, I was fated to make a conscious decision to respire; that without this breath my dives would be impossible? Did you watch as I shot gallons of air through my organ-valve nostrils, emphatically, each cloudy discharge creating a rainbow beneath the Sun? Again and again I would charge my body with Oxygen, until my lungs collapsed, folding among my ribs, depleted. With no air in my bones now, nor Nitrogen in my bloodstream, I'd be protected from the bends, a feat of engineering. Sounding a plosive *whoosh*, ambivalently I would fill my lungs entirely, ready to exercise my transformation. It was only then, my shiny blowhole closing like a trap, that I dove.

I am sure: you must have seen the placenta, and the teats gorged with milk. You experienced the great opening up of the womb, the live births, the suckling of the calves, and the motherly instruction. But did you ever contemplate, as you calmed yourself when night fell, that I could never sleep within you just like everyone else? Was your peace disturbed knowing that I would always need one half-brain awake? And tell me fairly, did you feel a tinge of guilt?

The Scientists wonder: Why do whales breach? Some say we are trying to dislodge parasites, for the force of our lunges is enough to slough off skin. Others have it that we're advertising our prowess, signaling our presence to contenders and mates. Still others offer that it is for fun that we leap. But none knows us, honestly, for I will tell you the truth of it: remembering our origins, we vault to escape you. If only for an instant, we yearn to surrender once more to Gravity. That is why we breach.

6

Sacrifice has become a synonym for moral highness. More often it is a mere form of blindness.

They call me Cotylocara, and I died 28 million years ago. I gained the world, unwittingly, back in the Eocene.

But I also lost my ground.

MEMORY

THE BEGINNINGS OF CONSCIOUSNESS

1

I can recall everything, like Funes the Memorious:

> The texture of a crevice in a rock I have only touched once.
> The face of a stranger whom I will either like or dislike.
> The place I found food three leagues from here three months ago.
> The shape of a cloud in the south at dawn.
> Every difference I have ever seen.
> My sentiments about my feelings.

I may have only 500 million neurons as compared to your 86 billion. Yet I have no blind spot. Remember that.

I have three hearts, and my blood is bluish-green.
But I am here to speak of other marvels.

2

I can remember how it happened, when the feeling first crept up stealthily, invading like a thief, only one who comes to give rather than plunder. It was long ago, too long for recollection. Still, mysteriously, it left a track.

Hundreds of millions of years ago an ancestor of ours was made of just one cell, wafting in the waters dimly. But through successive transformations it became two cells, and three, and four, and then plenty. Then the signals it formerly used to learn the outer world turned inward, to allow the different cells to talk to one another, itself to communicate with itself for the very first time. It was a strange sensation, like white noise, perhaps, but not much more.

An internal map was being laid down by neurons.
But there was no "I" yet of which to speak.

3

There were reflexes, of course, on which to count for survival. Our jellyfish progenitor would withdraw when touched and explore for food when there was none. It would nurse bodily damage and reproduce. In the absence of identity, there was an inner homeostasis that kept it alive, before it divined itself, for it strove to return to that place, perennially, evading or rectifying all uninvited departures.

This was very long ago. Do you remember, human?

4

Our deep forefathers started moving then, earnestly, to avoid salts and unpleasant temperatures, to satisfy a growing hunger or shy away from a threatening gale. That is how they began to learn, though I will be the first to admit: this was no extravagant education.

The principle was simple: the neurons that fired together wired together, leaving behind a heap of orphaned synapses. It was an unsentimental affair, but consequential, for our forefathers could modify their behavior now, and when I think of this it makes me glad.

Then, abruptly, our paths diverged.

5

It was memory that made what came next possible, on my side of things at least. For to learn anything well, my forebear had to memorize a stimulus-response relation and, when the time came, recall it. Only in this matter could it regularly avoid departures from the blissful inner state.

What happened on your branch I cannot say, with your central nervous systems and spinal cords and all their accoutrements. But it was a sensory signature that gave birth to my ancestor's "remembered present," and before too long it became its second nature. Thus, gradually, did incipient I improve at moving deliberately toward a food source, or with greater stealth away from a predator.

I say "incipient," for there was something now that it felt like to be oneself, a glimmer of subjective experience. I cannot truthfully describe it, nor precisely convey its meaning. Perhaps it was primordial emotions. But these were too fleeting and, in any case, too formless to celebrate.

6

Like you, I had been a Ctenophore first, then a Cnidarian, all those hundreds of millions of years ago when the drama unfolded.

But after we parted, the neural highways of my ancestors gave birth to a novelty of momentous proportions. Suddenly there was more than just pure physicality in the world to contemplate.

After more than 13 billion years, obscurely, matter created self.

7

I don't pretend these were anything but beginnings: the feeling of being oneself was still hopelessly amorphous. But then Oxygen levels reached a threshold, courtesy of the workings of photosynthesis. And as they provided the fuel for organisms to grow, including their expensive neural networks, life spans expanded, rendering new kinds of memory important. By now my mollusk progenitor had morphed into a kind of limpet, crawling on the seafloor. And at present it was ready to graduate.

For as the limpet grew, so did the appetites of its neighbors, until the Ediacaran transformed into the Cambrian and creatures concerned with no one were replaced by animals on the hunt. For the first time, mind would evolve in relation to other minds, and there could be no objections or exceptions or stunts.

8

In the beginning my limpet forebear had no inkling of a yesterday, only a precarious, inch-by-inch, everyday present. But stimulus and response were gradually decoupled, and memory tracks suddenly created a personal diary. Alongside its "remembered present," a "remembered future" now determined: one's history could begin to build one's character, gradually.

Before this happened my forebear would tremble at the sight of an Anomalocaris, and because it trembled, it feared. But thanks to the kind of memory it had engraved in its prey's nervous system, even when no ferocious Anomalocaris was present, my ancestor would now tremble because it first feared. This and other sentiments had arrived due to tricks of temporality.

In this world it was the possibility of a future that gave birth to the past.

9

Perhaps it was no accident that it went both ways, however. That when a past was born, the future, too, became more ever-present. And so, when associative learning arrived on the scene, and sensory states became fully fledged feelings, my ancestor gained a bigger shell to cover itself, protection from a hostile environment. And just as the passions seeped into its being, gradually filling its palate with annoyance and contentment and thrill, just then so did gas bubbles make their way between its shell and body, tickling, hinting at the opportunity for greater things. Perhaps sensing hope, the limpet rose into the water column, detaching its foot from the seafloor, a free agent. Its growing shell had become a floating device, a cone-shaped liberator.

And the great blue above whispered to him, *Follow.*

10

You look at me now and I wonder:

Do you want to eat me?
Do you want to play?

I peer into your eyes, but I cannot see.

How much are you alive?
What are you thinking?
Are you happy?

If you are wondering, my penis is detachable, and I can see you with my skin. But let me first relate more wonders, tell you where I've been.

11

It was before the dinosaurs arrived on land that my ancestor decided: Enough! Its shell had helped it rise, but it was time for independence. For its inner world was now connected to the outer one, experiencing not just sensing, intending rather than duly exercising the rites of rote survival. And so a second, more familial divergence came about, 320 million years ago. On one side the mollusks abandoned their shells entirely, and on the other they swallowed them, gaining a mantle within. The latter became my cousins, the squid and cuttlefish, and the former my direct progenitors (only our dim relative, the nautilus, remained within its shell, unenlightened). In all of us cephalopods a siphon was now marshaled to fashion a water jet, a new trick for movement in lieu of our abandoned, or swallowed, shells. Ink sacs for squirting, too, were born for the purpose of diversion. We would now be on our own in this Shangri-La/Hell.

It was then that the most dramatic metamorphosis occurred, at least in my remembered pedigree. Having found itself profligate, and evolution's creative mind gone on the prowl, the foot that had once anchored the limpet to the seafloor morphed into eight separate tentacles, each with 1,600 suction cups and 10,000 "taste" receptors. This is when I became myself, finally.

12

It is true, I cannot abstract well, because I cannot forget a difference. And when I hatch, I drift away, solitary, and that is how I like it. But please, long-living creature, come closer, don't be shy, tell me: Are you infused by thought or controlled by it? When you breathe do you contemplate breathing, or does it just happen, automatically?

Body. Mind. Your distinctions are meaningless to me: only two-fifths of my neurons are in my head—the rest are spread among my arms, autonomous. You call it "distributed intelligence," and I say: I am both self and other, and neither self nor other—I have no boundary.

I am everywhere and nowhere.

13

Look and see, when the big waves come, I will disappear to a cave within the waters, holding on to the crags and not letting go, like my ancestor limpets did within their shell, until their storms passed, long before me.

But beware: even when it's calm, I can vanish before your very eyes, just like that, in an instant. At a whim I am a kaleidoscope, perfectly matching iridescent colors to my background. Before you know it, against a coral reef or kelp woodland, I am gone, though ever-present. For below my skin there are tiny color sacs, and below them, more-luminous reflectors. You should not be surprised at your Scientists' discovery that I control my exterior with the very neuromolecule that effects what you call arousal. Camouflage began as a way to hide from others but soon evolved into a form of disclosure.

I show it in my patterns: there are times when I seek your company, and others when your presence can only make me moan. I cannot tell you why.

Now go away.
I wish to be alone.

14

Come back, human.
Give me your hand.
Let me taste it.

I want to share my world with you.
But you would not be able to understand me.

Even if I could speak.

TRUTH

LANGUAGE IS BORN

1

They could use their eyes, with all their expression. And like the octopus, their colors. They could reveal teeth, raise hairs on their backs, point a tail, perform a dance. There were chemicals to mobilize, and odors to marshal, and poses to strike, of great meaning. And, of course, there were also the sounds: the buzz, the click, the shriek, the howl, the croak, the bellow, the bleat, the hiss, the bark, the moo, the oink, the roar, the whine, the yap, the bray, the honk, the purr, the screech, the cluck, the crow, the chirp, the creak, the bell, the coo, the growl, the whistle, the hum, the chirrup, the tweet, the twitter, the trumpet.

All these the animals could use, and get across a message. But they could not lie, for there was no truth yet to violate. Lying would need to wait for the "upright man."

2

It was Homo erectus who began the great tearing-away of the hominins, long ago on the plains of the Pleistocene. It was he who first tamed the mysterious fire.

And as the unruly flames were turned into a hissing, purring pet, so was the brute himself domesticated, in a feedback loop of momentous occasion. For by cooking his food now, Homo erectus made it more digestible, allowing his jaws and teeth and guts to shrink, and gaining more calories to fuel his intellect. And as the fire became the communal campground, his brain expanded within his cranium; as camaraderie was invented, and table manners, and decorum, so gradually was the unshared life abandoned, and the lure and grasp of its savage ways disavowed.

Remaining together in his band more readily, Homo erectus became increasingly dependent on the others—for sustenance, and protection, and with time his very happiness. It was then that apprenticeship and pedagogy were born, instruction in tool use, cooperative hunting, and the collective forage. Alloparenting appeared, and care for the infirm, even the demented. Helping the group to bond more easily, rituals were contrived, forging a common identity. And as art was produced, an appreciation for beauty was cultivated. The ape was slowly turning into a man.

And so perhaps it has been a misnomer: *Homo gastronomicus* would be more fitting than *erectus*. Culture and biology remained wedded, despite civilizations' future self-congratulation. There was no "graduation" from the body, no magic disembodiment, or Rubicon crossed—just a campfire. Humans have become accustomed to believing they created technology. Instead, it was technology that created them.

3

It is true, an ape could follow a fellow ape's gaze, attributing beliefs and desires, plotting stratagems. But as Homo erectus sat cooking beside the fire, something strange happened, for the first time, to animals: their survival became dependent on peering into each other's eyes. Now it was suddenly worthwhile to take the time to wonder, deliberately: What is it that you *really* think? How do you feel? and Why? It was the deep gaze, not the fleeting glance, that would make all the difference—the mutual quest for understanding, not the hustle for the private edge. This would hasten the great tearing-away of the hominins, their gradual departure from the wild.

And so, slowly, Homo erectus learned to read his neighbor. Catch him blushing. Perceive how his lips curl. Notice the inkling of a furrowed brow. And as they came to know each other more intimately, the former brutes began to pair-bond and show affection, to empathize and sympathize, to judge and beguile.

Soon Homo erectus could point at a flower, or a scheming predator, depict the world to a friend. He could smile at the elation on the face of his bandmate, crowding beneath a frond, with the tympany of first rain above. And when sorrow came at the loss of a loved one, he could commiserate, lend a shoulder, standing beside the hardening corpse in league.

It was mimesis that was the missing link between ape talk and human language, mimesis that bridged the gap between one searching mind and the other. For living together in proximity and watching each other intently, hominins could begin to imitate one another and, with time, develop traditions and folklore. The stars were out there, the Moon and Sun and planets and the great expanse. The rivers and rocks and trees and ravines. The smell of dusk and the howling of wolves and the wind's caress—all were there to experience together. And slowly the group united, becoming a communion of mind.

4

But there was a limit to mimesis, like a wall of vast elevation, blocking the view to greater things. For to participate in experience, it would need to be shared, literally, in person: the winds and howls and aromas, the rocks and ravines, the rivers and trees, the stiffening body and dancing frond, the firmament and the great big expanse—all would have to be there, before the senses, to contemplate. Thus God and infinity were precluded.

Necessity, it is said, is the mother of invention and, despite the exceptions, for good reason. For if Homo erectus, alone, had seen a lion in the bush, what would he do? With no mates in his vicinity to point out the beast to, how could he inform the band of the lurking peril? And if, crafting a hand ax, it occurred to him that one could use it not only to cut shanks, like the ones here before them, but to shape other rocks like the ones he had seen when he was alone by the river downstream—how would he announce the flash of insight to his comrades? How could he give shape to a peculiar notion lurking nowhere but within him? Mimesis would not be enough.

And so, by compulsion, Homo erectus began to build a new technology. And like the fire, it too would change him profoundly. There are those who still believe that there was a special organ inside the brain for this, but it was a social tool, not a blade on the brain's Swiss Army knife, that would usher in the revolution.

Today we call it Language.

5

It is true, animals could discern danger and call out to each other to avoid it: the chipmunk cries for hawks and foxes differently, and monkeys have separate hollers for eagles, leopards, and snakes. Even the plants, Darwin knew, can merrily communicate, sending distinct underground missives through chemicals in their roots.

But despite the various tweets and howls and roars and yaps and grunts and purrs and diffusions—despite all these and many other memoranda, messages on Earth had been linked to direct experience, exclusively. Then came Homo erectus and forged another track.

As speech evolved as a communal code, it began to be used in a more intimate way—to dream the world, not just brand it. Soon there was more to life than merely the apparent: a new realm, of thoughts about the absent, feelings on what lay beyond the senses in the mundane here and now. Promises could now be contracted, and futures explored. Gossip could be used to build and break allegiances. Above all, minds and hearts would open up, a millionfold: What did it feel like to mistake a coming storm for a passing drizzle, and be trapped on the outcrop with nowhere to go? Or to watch the color blue glide gently into orange above the ocean, then purple, then turn to darkness, and then light again, in the morning glow? Even when the subjects of his thoughts were not apparent, Homo erectus could now begin to share his private musings. And thus, using language to instruct his friends' imagination through his own experience, his world and theirs began to grow.

Soon there were reasons to trust a friend, more or less, to take his word for things, even when those things were invisible. The crux of the revolution was the will to follow the imagination of another, wherever it would lead. And so gradually, a kind of double mind was born in our ancestors: the first producing inner experience—private and subjective—the second a social outlook, helping individuals become part of their group. In the attempt to align the two, new kinds of understanding appeared and, with them, new confusions. A few of the animals had this duality in precious rudiment. But with adjustments the recursive symbols began to do their magic.

And for the first time, there was Truth.

6

Language brought Truth, but also its corollary. For, thanks to the symbols, there was suddenly far greater opportunity to depart from the facts than to represent them—just one meager reality to delineate, but an infinitude to imagine. Thus, while the other animals merely deceived each other in piddling, petty ways, Homo erectus began distinguishing Truth from Falsehood. Necessity, they say, is the mother of invention, but often they are sorely mistaken. For the Lie did not appear to solve a private or communal problem, at least not originally. Having arrived, instead, it was put to use.

Many Scientists would later claim that to safeguard speech, early hominins would need to rein in Falsehood, thinking that Truth would otherwise collapse entirely, and with it the group's survival. But once it came, the Lie could not be constrained, nor would it have to be; in many different ways it was superior to reality. Thus, the white lie, the exaggeration, the half-truth, the fib, the perjury, the promise, the myth were all invented—and the brains that would willingly or unwillingly be deceived. On the wings of imagination, fiction was born, in all its duplicitous splendor. And with it the contemplation of all possible things.

7

Language brought Truth, some even say the very Universe. But it was the Lie in the end that would make us fully human. For the Lie, not the Truth, would hasten the departure from the Garden of Eden.

HOPE

TRICKS OF THE MIND

$w\Delta z = cov(w_i, z_i)$

1

And so every story finally returns to its beginning.

I was seven years old when I first encountered *The World of Myth and Legend*, a large and colorful volume illustrated by Rob McCaig from Brimax Books. Opening its hard cover as one would a vault, I fell upon Theseus outwitting the Minotaur with the help of Ariadne's ball of string. There he was, elated on his voyage home to Athens, with his lover on deck beside him, only to forget his promise to his father to hoist the red sail as a sign of victory. And the King, seeing the black sail instead, and therefore certain of his son's death at the hands of the monster, throwing himself off the cliff into the sea that would one day bear his name.

I held my breath when, wary of King Minos's wrath, the architect of the Minotaur's labyrinth, Daedalus, quickly fastened white wings to his son's adolescent back with wax, warning him as they escaped the island not to fly too close to the Sun. And I wept on my pillow as he watched Icarus plunge into the same watery graveyard that King Aegeus had fallen into, a victim of his own euphoria. Why had Icarus not heeded his father's words? And why had Theseus forgotten to raise the red flag? The answers did not come.

The myths had invaded my dreams. I thought of them constantly. I could hardly believe the courage of the mighty Green Knight of the North, marching on his green horse at Merlin's behest into Camelot to challenge King Arthur's Knights of the Round Table. For long hours I stared at the blade of his giant ax, which would fell his own head with one blow and promise to do likewise to Sir Gawain, *quid pro quo*, the next Christmastide. My mind raced: How had he mended the head that had fallen? And how, to test Sir Gawain's resolve, had he morphed into a nobleman, then a huntsman, and finally a knight, all three of them peddling distractions on Sir Gawain's journey to a morbid destiny? Finishing the tale, I stared at the headless platter and wondered: Would my destiny, too, one day be surprisingly altered? Then I peered again into the silent darkness.

Beneath a sheaf of light from the blue lamp affixed to my little bed's headboard, I flipped a few pages backward. Thor's great hammer, Mjölnir, had been hurled and was about to smash the giant Rungnir's forehead. How had the cunning Loki made the earth tremble beneath Rungnir's feet, bringing mead and merriness back to Valhalla? Why did the gods beat the giants, despite it all?

I pondered and puzzled, but I did not know.

2

In high school my teachers taught me Euclid's geometry, the rules of covalent bonds, the Second Law of Thermodynamics. My heart broke then for the first time, in the pubescent halls between the classrooms, and I discovered to my chagrin the putrid poison of jealousy. But I remember, too, peering down the ocular of a microscope, how I first saw a cell, and within it a mitochondrion. I gasped, unbelieving. And my heart was full again.

Later, I swam with an orca in the Sea of Cortez, weeping underwater for a brother I had lost in battle. How had chance spared me and taken him instead? And was the sacrifice worth it? I did not know. Nor could I mend.

One summer I spent with cuttlefish, near the ocean, scoring how they teach one another to eat a crab. The next I was behind a one-way window in the anteroom, parents fidgeting nervously beside me, as a psychologist in the theater administered the Vineland Test to their toddlers, the mothers and fathers mouthing the answers, prodding with frightened encouragement from afar, awaiting verdicts. And then again in the laboratory at that storied institution, searching out the genes for schizophrenia and OCD.

It was at that time that I learned about infinity: that the points on a line between 0 and 1 are a larger quantity than of all the natural numbers; that not all infinities are equal, and there are ones so big they can swallow up the others. How could this be? I puzzled and I pondered. And when I thought I knew the answer for fleeting moments of blessed clarity, the insight slipped between my fingers, like fleeing grains of sand.

3

With pride I marched into the university to drink up knowledge. I learned the histories of al-Khwarizmi, the Persian astronomer, inventing the rules that would lead to our machine algorithms, and Steno, the Danish geologist, reading the Earth like a palimpsest. I imbibed the spirit of the Enlightenment, and tried getting into the heads of the Luddites as much as I could. There was Maxwell sending electromagnetic waves through the ether, and Michelson and Morley showing that the ether was make-believe. There were the Germans Schleiden and Schwann and Virchow and their cell theory; Tesla the inventor; Haber the yields multiplier; and Élie Metchnikoff, the immune system man, fighting off disease. And then there were the ones who went straight to the heart: the American outcast who tried to solve the riddle of altruism, and the self-taught boy from Madras, a genius with numbers, who awed the dons of Cambridge, and then expired.

Carefully I read the philosophers, on dualism and reduction, God and the Infinite, metaphysics, language, and the limits of the mind. Aristotle, Lucretius, Maimonides, and Vyasa; Pico della Mirandola (what a name!), Leibniz, and Spinoza; Locke and Hume, Bishop Berkeley, and Kant. I read the utopias of Sir Thomas More and Sir Francis Bacon, embraced Condorcet, that doomed French reformer, and the positivism of his countryman Auguste Comte. Fumbling from one to the next, I learned of the hope to rise above Nature, on the wings of reason, finally.

But I also met Schopenhauer, Zamyatin, and Nietzsche. I trudged through Gödel's Incompleteness Theorem and was astounded by Kierkegaard's challenge—"I believe therefore I am." In and out and in again I fell with the troubled Wittgenstein. At a museum one rainy day, I was spellbound by the rectangles of Rothko and later, in a concert hall, Jean Sibelius's violin concerto. I lived more, and saw more, and fell in love and had children. And just as with Icarus, the higher the modern world flew, the more quickly did it seem to me that its wings were also melting. Certainly, there were longer life spans and fewer wars now, rights for outcasts, global travel, knowledge at the fingertips, the Internet of Things. But there were also the weapons of mass destruction, fundamentalism, gross inequalities and ecological disasters, government and corporate surveillance, refugees, and continued spread of disease. The "Risk Society" had produced results even it could not anticipate. And so a kind of incredulity began to seep through the cracks of deliverance: Will the march of progress solve all quandaries? Will it really deliver on all of our dreams?

Always, I kept the myths close to me, clutching to thoughts beyond our provenance. For it seemed there was even greater solace now in the myths' inscrutability. And greater wisdom in their schemes.

4

In the summers when I was a boy we would come to visit my grandmother in New York City. My older sister would get so excited: Broadway plays, tchotchke shops, restaurants, the bookstores. But for me it was the American Museum of Natural History on Central Park West and 79th Street that would get my heart thumping, climbing two stairs a pop below the bronzed president, Theodore Roosevelt, mounted on his massive steed. When I walked through the halls, the entire world faded away behind me, the echoes from the vaults becoming muted as if we were underwater, just me and the fossils communing. I looked up at the pterosaur—what a funny crest and beak it had, and that fourth finger, elongated, to scaffold its majestic glide. I quickened my step as the blue glint of the Hall of Ocean Life became visible, knowing that I would soon see my old friend the great whale, suspended in midair before me. Later I would stand glued before the squat couple marching out of Africa, the man's furry hand on the woman's furry shoulder, leaving behind their 3-million-year-old tracks. Could these really be my ancestors, my grandparents' progenitors? For hours I stared. It seemed as mystical to me then as a stellar parallax.

I can remember the day the coin dropped, in the form of Darwin's currency—that we had evolved, from simple beginnings. That our appendix had once helped a fish float, our tailbone a monkey swing; that our goose bumps pricked up hairs, long ago, to scare off ancient predators. The wisdom teeth crowding our modern mouths and the pseudogenes of olfaction all told the same tale: there was a history; we hadn't come from nowhere. Before we were us, we were different but similar. And before that, we were some sort of snail.

Gradually, I learned the story. That Archaea and Bacteria preceded us, and before them, possibly, the Ribozyme. That our fate rolled precariously through a pilus, back and forth. That slowly, lacking foresight, our branch had begun climbing—upward and sideways, to the left and right, thicket-like. That, forsaking autonomy, cells had gathered to create complex life-forms; that sex was born, frantically, thanks to a micron-long nervous symbiont; that although it was evolution's handmaiden, sex would lead, inevitably, to our own demise. And a body had gained a neural network, and a neural network had gained a memory, and a memory had forged a language, and a language a brain with a spirit, whatever that may mean. Onto land life crawled from the seacoast; then some creatures vaulted to the heavens, and others dove back into the waters, like returning ghosts. We had not been the first either to covet or to sacrifice, to communicate, fly, or cooperate. But somehow, from Methanogen through Trilobite, tetrapod through hominin, we arrived, walking on our feet.

Huddled around their fires, our ancestors were transformed. In order to live together amicably they had no choice but to develop trust. Thus was guilt born, and—as the numbers grew—punishment. Empathy, too, and a sense of communion. Slowly the ego made way for the other, in expanding circles, like ripples in a pool. And as the genes and hormones and physiologies and anatomies evolved, in eternal feedback with culture, the great metamorphoses came dramatically to pass: cognition birthed discourse; togetherness brought mind reading; empathy, morality; memory, the possibility of tomorrow. With time, philosophy was born, and—on the wings of imagination—literatures and poetry. Technologies proliferated to provide the glue to stick together: script and law, religion and currency. Soon the tribe became the village, the city, the nation, the globe. And the little lies of livelihood morphed into grand collective stories, the ones we would one day call mythologies.

5

On Friday afternoons I would walk to a cold spring in the hills of Jerusalem. Sitting in the freezing waters beneath cypress and pine trees, I listened to the silence of the Sabbath descend on the world, and my thoughts would wander.

Hadn't I learned yet? Our planet revolves around a sun that is one of billions of stars in our galaxy. The 100 billion known galaxies are just a fraction of those in the Universe, and our 13.799 ± 0.021-year-old universe is possibly just one of an infinite number of universes, each one a bubble, destined never to meet one another in the plenitude of dimensions. Did I not already know that nearly 4 billion years ago life emerged on Earth from prebiotic chemicals; that we are a twig on one of its branches, the descendants of a single species of African primate; that we developed agriculture, government, and writing late in our history, and romantic love just moments before our present?

Be honest, I'd tell myself. Aren't the beliefs of all the world's religions and cultures factually mistaken, their theories of origins nothing but childish fictions? Should not prophecy and revelation, dogma and tradition, be dismissed as sources of knowledge, much less of truth?

I would feel the warm wind on my ears, the sun caressing me as it prepared to set. But there was a storm raging inside me. Had it not been established that on scales very large and very small our intuitions trick us, about space and matter and time and causality? That the discrepancy between the laws of probability and the workings of cognition helps explain our suspicion of coincidence, why we put faith in omens and curses, retribution and prayers? Had we not learned that we are not rational, despite it all, and depart from rationality in ways that are predictable? That rationality is in any case just one of many ways to score experience? That we can only know the world through the senses that evolution has passed down to us; that reality is therefore forever opaque to us, our relationship to truth an asymptote, like unrequited love?

I had trained in the habit of conjectures and refutations. And so I knew that regardless of convictions, what we hold true today would surely crumble tomorrow. Thus science, too, was an authority on a stopwatch. This is the way it should be, I'd calm myself. There is more to life, anyway, than we can know.

In the final reckoning, we humans are hybrids, like the fire-breathing chimera of Lycia, walking the banks of our lives. For unlike the ants, we haven't completed our transition, the collective never conquering the I, at least not yet. Part selfish, part benevolent, our kind is in turns cruel and chivalrous, full of both trust and suspicion. Stuck halfway between independence and integration, we are an ambivalent superorganism, the envy of ants, and their laughingstock. Thus, like Janus, we look in both directions. It was fitting the Romans had made him god of beginnings and endings, doorways and passageways and gates. This was nature's revenge, our mixed-blessing inheritance: infinity is boundless, but we are finite.

Torn between inner and outer worlds, freedom and belonging, group and self, logic and absurdity, meaning and insignificance, we march along with hope. I know: the laws governing the physical world, including accidents, disease, and luck, have no goals, good or bad, for us. Providence and Fate are mere inventions.

And so, I'll hold tight to my myths, as solace against all of this. I will bow to science watchfully but take heed of precious legends. And as I march, I too will trip and fumble and fail to learn from experience. I'll soar and plunge and love and sacrifice, utterly despair and once again find faith. More than a few times, I'll succumb to hubris, maybe gluttony and avarice. All this I will do, and cultivate inquisitiveness. For I am a human, and like you, I remember the past as I dream of immortality. In a universe that does not care for me, one day I too shall pass.

6

When I put my kids to bed, I tell them myths and stories. They are young and their wonder is alive. Beneath a sheaf of light from a blue lamp just like my old one, I flip the pages of precious books, in the silence of the night.

And just as their eyes are closing, their little fingers curled around my own; just as they are slipping into their world of dreams, I whisper . . .

Remember, my loves, and try to make peace with it:
These things are never, and are always.

ILLUMINATIONS

The world of myth is both wondrous and vast; it would be almost impossible to cite all of the sources that have informed *Evolutions*. Still, I have had some essential guides in the caverns of world mythology: William G. Doty's *Mythography: The Study of Myths and Rituals* (Alabama University Press, 2000) is one; G. S. Kirk's older *Myth: Its Meaning and Functions in Ancient and Other Cultures* (Cambridge University Press, and University of California Press, 1970) is another; Karen Armstrong's *A Short History of Myth* (Canongate, 2005) is a third; and Umberto Eco's *Storia delle Terre e dei Luoghi Leggendari* (Bompiani, 2013), which I read in Hebrew, is yet another. So too have six classics been indispensable to my thinking: Joseph Campbell's *The Hero with a Thousand Faces* (Bollingen Foundation/Pantheon Books, 1949), Bronislaw Malinowski's *Magic, Science and Religion and Other Essays* (The Free Press, 1948), Mircea Eliade's *Gods, Goddesses, and Myths of Creation* (Harper and Row, 1974), Bruno Snell's *The Discovery of the Mind* (Harvard University Press, 1963), Claude Lévi-Strauss's *The Raw and the Cooked: Introduction to a Science of Mythology, Vol. 1* (1964; Harper and Row, 1969), and James Frazer's *The Golden Bough* (1890; new abridged ed., Oxford University Press, 1994). These monumental works have been the subject of much debate and even criticism, but they remain touchstones for any consideration of myth.

As a child I could not get enough of the beautifully illustrated book *The World of Myth and Legend* (Brimax Books, 1980), Edith Hamilton's classic *Mythology: Timeless Tales of Gods and Heroes* (Little, Brown, 1942; Back Bay, 1998), and Ovid's *Metamorphoses*, as well as the Old Testament of the Bible, especially the stories of Genesis. Later I discovered that Penguin Classics has a series of annotated world mythologies, from *Hindu Myths* to *The Mabinogion* to *Tales from the Thousand and One Nights* and more. Alongside the classics, current writers including Chinua Achebe, Margaret Atwood, David Grossman, A. S. Byatt, and Jeanette Winterson provided modern renditions of old myths in a beautiful series commissioned by Canongate. So, too, for the Greek tradition, have Ted Hughes and Roberto Calasso each in his own way written modern renditions of classic stories, in *Tales from Ovid* (Farrar, Straus and Giroux, 1997) and *The Marriage of Cadmus and Harmony* (Knopf, 1993), respectively. Italo Calvino struck out on his own path, in the wonderful *Cosmicomics*. *De Rerum Natura*, written in the first century B.C. by the Roman Epicurean Lucretius, taught me endlessly about how to view the universe as a poem. Darwin's colorful grandfather, Erasmus, especially in his *The Temple of Nature; or, The Origin of Society* (J. Johnson, 1803), made me laugh and smile. These and many other stories, old and new, have served as inspiration.

Evolutions concerns the truth claims of science, and here Paul Feyerabend, a dead Austrian philosopher with whom I argue in my head on many matters, was a gadfly. In his book *Against Method: Outline of an Anarchistic Theory of Knowledge* (New Left Books, 1975; Verso, 4th ed., 2010), Feyerabend claimed that while scientists and philosophers may want to draw a strict separation between scientific and nonscientific forms of knowledge, in actual fact such a separation is a fantasy: the history of science shows that there is no universal, rational method by which science advances. What this meant to certain of his readers, and to many people's shock and horror, is that science cannot be judged better than alternative, and incommensurable, worldviews, like mythologies. In *Philosophy of Nature* (Polity Press, 2016), discovered posthumously among unpublished papers, Feyerabend returns to the caveman and deep antiquity, from beyond the grave, to show why this is true.

The myths in this book draw from the science of our day, beginning at the dawn of time. There are many general-audience books both on the birth of the universe and on the history of the solar system. Philip Morrison's book *Powers of Ten: About the Relative Size of Things in the Universe* (Scientific American Books, 1990) provides mind-bending perspective, to begin with, taking readers on a magical journey from a billion light-years away to within the realm of the atom. Carl Sagan's *Cosmos* (Random House, 1980), based on his famous television series from the 1970s, is a classic but dated; David J. Eicher brings us up to speed in *The New Cosmos: Answering Astronomy's Big Questions* (Cambridge University Press, 2016).

Lisa Randall's book *Dark Matter and the Dinosaurs: The Astounding Interconnectedness of the Universe* (Ecco, 2015) is a fascinating attempt to link cosmology with life on earth. More recently, Sean Carroll offers his own thoughts on the matter in *The Big Picture: On the Origins of Life, Meaning, and the Universe Itself* (Dutton, 2016). Randall and Carroll are both physicist-astronomers. Among the many excellent books on the evolution of life on the planet written by biologists, Nick Lane's *Life Ascending: Ten Great Inventions of Evolution* (W. W. Norton, 2009) stands out for its clarity and thought-provoking suggestions as a "greatest hits" version of evolution, as does his more recent and metabolically focused *The Vital Question: Energy, Evolution, and the Origin of Complex Life* (W. W. Norton, 2015).

More than twenty years ago John Maynard Smith and Eörs Szathmáry's book *The Major Transitions in Evolution* (Oxford University Press, 1995) spurred a great deal of discussion on the idea that the evolution of life can be viewed as an evolution in the way information is stored and transmitted. This idea has been returned to in *The Major Transitions in Evolution Revisited*, edited by Brett Calcott and Kim Sterelny (MIT Press, 2011), in which the focus becomes life as a series of transitions in individuality, from the individuality of the self-replicating molecule through the individuality of the chromosome, the genome, the organism, the group, the lineage. The jury is still out on the topic of transitions, and both books are worth reading. Less theoretically ambitious but nevertheless highly informative, Peter Ward and Joe Kirschvink's *A New History of Life: The Radical New Discoveries about the Origins and Evolution of Life on Earth* (Bloomsbury, 2015) presents a complement and update to Richard Fortey's entertaining *Life: A Natural History of the First Four Billion Years of Life on Earth* (Knopf, 1997). Andrew H. Knoll's *Life on a Young Planet: The First Three Billion Years of Evolution on Earth* (Princeton University Press, 2003; updated ed., 2015) is also excellent, as is the new edition of Richard Dawkins and Yan Wong's Darwinian-centered account in *The Ancestor's Tale: A Pilgrimage to the Dawn of Evolution* (Mariner Books, 2016).

Scientists often ascribe agency to their molecules, organisms, even equations, though it is considered a big no-no and a categorical mistake. This was not always the case. For centuries, spirit and matter were thought of as inseparable, and agency was part and parcel of scientific descriptions of the physical and natural worlds. For a history of the attempt to ban agency from the life sciences, including the sustained voices of dissenters, see Jessica Riskin, *The Restless Clock: A History of the Centuries-Long Argument Over What Makes Living Things Tick* (University of Chicago Press, 2016).

In the bibliographical essays that follow, more-specific scientific references are provided, to books and articles, websites, and popular sources that have informed the writing of the myths, and that help to put them into context. While these are merely representative of a much larger set of studies, readers interested in expanding their knowledge can use them as a springboard from which to dive into deeper waters.

FATE

What hasn't been said about the universe? Immortal wisecracks include that it was dictated but not signed (Christopher Morley); that there's a coherent plan in the universe, but a plan for what precisely remains a mystery (Fred Hoyle); that it's astounding that people want to "know" the universe when it's hard enough to find your way around Chinatown (Woody Allen). One of my favorites is Douglas Adams: "There is a theory which states that if ever anyone discovers exactly what the universe is for and why it is here, it will instantly disappear and be replaced by something even more bizarre and inexplicable. There is another theory which states that this has already happened."

For most of human history, the origins and scope of the universe remained a mystery, a fact we have to thank for some of our most beautiful ancient myths. Recently, exciting developments in physics and astronomy have begun to shed their illuminating light. Using both theory and experiments, including ones performed at the largest experimental facility ever built, the CERN Large Hadron Collider near Geneva, scientists now feel confident that they can explain the forward evolution of the universe from what has been termed the big bang. Parsing precisely what happened at each moment, they have defined the

- Planck Epoch ($<10^{-43}$ seconds): dominated by quantum gravity effects

- Grand Unification Epoch (<10^{-36} sec): gravity peels away, the other three forces remain united, and the first elementary particles and antiparticles begin to be formed
- Inflationary Epoch (<10^{-32} sec): cosmic inflation expands space by a factor of 10^{26}; the universe is supercooled; the strong forces become distinct from the weak forces
- Quark Epoch (>10^{-12} sec): the forces have separated, but energies are too high for quarks to coalesce into hadrons
- Hadron Epoch (10^{-6} sec–1 sec): quarks bind into hadrons; antihadrons are eliminated
- Lepton Epoch (1 sec–10 sec): leptons and anti-leptons remain in thermal equilibrium; neutrinos decouple
- Big Bang Nucleosynthesis (10 sec–10^3 sec): protons and neutrons are bound into primordial atomic nuclei
- Photon Epoch (10^3 sec–380,000 years): the universe is a plasma of nuclei, electrons, and photons; temperatures are still too high for the binding of electrons to nuclei
- Recombination (380,000 years): electrons and atomic nuclei become bound to form neutral atoms; photons are no longer in thermal equilibrium with matter, and the universe becomes transparent; photons of cosmic microwave background radiation are formed
- Dark Age (380,000–150 million years): the time between recombination and formation of the first stars. From here on, the millions and billions of years that elapsed since the big bang are represented as mega-annum (Ma) and giga-annum (Ga), respectively.
- Reionization (150 Ma–1 Ga): earliest "modern" Population III stars are formed
- Galaxy formation and evolution (1 Ga–10 Ga): galaxies coalesce into clusters and superclusters
- Dark Energy–Dominated Era (>10 Ga): matter density falls beneath dark energy density, and the expansion of space begins to accelerate; the solar system is formed, and evolutionary life begins

To learn more about these amazing discoveries, treat yourself to the California Institute of Technology cosmologist Sean Carroll's two-part *Dark Matter, Dark Energy: The Dark Side of the Universe* in the series The

Great Courses, published by The Teaching Company, 2007. For the latest on the reconstruction of the trillionth of a trillionth of a trillionth of a second after the big bang, see "Shut Up and Measure: A Conversation with Brian G. Keating," *The Edge*, October 20, 2017.

Apart from how the universe evolved after the event, what came before the big bang itself, as most laymen intuit, remains shrouded in mystery. Was there anything before the explosion, and can something really come of nothing? These are questions for which educated theories exist, but also little certainty. We simply don't know the answer, and perhaps never will. To get a feel for the birth of the idea and its persisting obscurity, read Simon Singh's popular book *Big Bang: The Origin of the Universe* (Fourth Estate, 2004)—it's well written and fun. And to read more about the mystery of what led to the big bang and the expanding universe, read Delia Perlov and Alex Vilenkin, who has been a champion of the idea of something coming from nothing, *Cosmology for the Curious* (Springer, 2017).

That we should be here at all is a miracle, or so it seems. The "cosmological constant" is technically the value of the energy density of the vacuum of space, but what it actually means is that when the laws of physics are taken into consideration, the chances of us being here to contemplate our own navels are vanishingly small, practically nonexistent. One way to explain—some say escape—the daunting serendipity captured by the "cosmological constant" has been to posit a universe made of strings. For an excellent popular exposition of string theory, see Brian Greene, *The Elegant Universe: Superstrings, Hidden Dimensions, and the Quest for the Ultimate Theory* (W. W. Norton, 2004; new edition, 2010). Greene also explains why string theory implies multiverses in *The Hidden Reality: Parallel Universes and the Deep Laws of the Cosmos* (Knopf, 2011), but it's important to remember that string theory and its multiverse correlate are controversial: they remain in large part an elegant but unproven mathematical solution to a real-world problem—maybe the hardest problem we know, alongside the human brain and consciousness. Speculative cosmological theories don't have a great track record. The physicist John Archibald Wheeler used to recall a colleague's advice: "Never run after a bus, a member of the opposite sex, or a cosmological theory—because there'll always be another one in a few minutes." For an introduction to some of the criticisms of string theory, there is Peter Woit's *Not Even*

Wrong: The Failure of String Theory and the Search for Unity in Physical Law (Basic, 2006).

For all we think we know, cosmology, perhaps as much as any other field in science, is changing at an ever-quickening pace. For a wry popular-historical account of humankind's attempts to pry open the secrets of the heavens, you can't beat Arthur Koestler's classic, *The Sleepwalkers: A History of Man's Changing Vision of the Universe* (Macmillan, 1959). On the historical role of Edwin Hubble, read Gale E. Christianson, *Edwin Hubble: Mariner of the Nebulae* (Farrar, Straus and Giroux, 1995), and on a broader cast of characters, including Hubble but also Einstein, Lemaitre, Eddington, and others, see Marcia Bartusiak, *The Day We Found the Universe* (Pantheon, 2009). For an entertaining, globe-trotting look at astronomers' exploits today, read the science journalist Anil Ananthaswamy's *The Edge of Physics: A Journey to Earth's Extremes to Unlock the Secrets of the Universe* (Houghton Mifflin Harcourt, 2010). It will reveal just how much remains unknown.

HUBRIS

The Buddha said: "Three things cannot be long hidden: the sun, and moon, and the truth." Perhaps. But it sure did take a while to figure out that the earth isn't the center of the universe, and, once we learned the earth revolves around the sun, that the sun isn't the center of all things, either. A broad and entertaining literary-historical book about the role of our sun in mythology, language, religion, science, art, politics, and medicine is Richard Cohen's *Chasing the Sun: The Epic Story of the Star That Gives Us Life* (Random, 2010).

Most people think it was Copernicus who first claimed that the earth orbits the sun, but they're mistaken: the ancient Greek astronomer Aristarchus of Samos is the originator of the heliocentric hypothesis. When it finally supplanted earth-centrism in the seventeenth century, the solar system was still thought of as central in the universe; giving up our special place in creation would not be easy. Amazingly, the removed location of the sun in our galaxy wasn't actually worked out until centuries later, in 1993. See M. J. Reid, "The Distance to the Center of the Galaxy," *Annual Review of Astronomy and Astrophysics* 31:345–72, 1993, to experience the excitement (or diminution). Subsequent measures are even more precise.

Thoreau thought he had it right, gazing from the shores of his pond, when he wrote in the final sentences of *Walden*: "The light which puts out our eyes is darkness to us. Only that day dawns to which we are awake. There is more day to dawn. The sun is but a morning star." In truth, we have learned a great deal from science about the origin of our own light source. For a comprehensive astronomical account of the origin and evolution of the solar system, including discussions of all the planets (and the dwarf planet Pluto), read Michael M. Woolfson, *The Origin and Evolution of the Solar System* (Institute of Physics Publishing, 2000).

What the sun is made of, how it came to be, and where it's going are less mysterious than ever. Still, a recent discovery gives room for pause. A group of researchers at Caltech has suggested the probable existence of a mysterious ninth planet in the solar system. Elizabeth Bailey, Mike Brown, and Konstantin Batygin claim that the planet is responsible for the sun's dramatic 6-degree tilt relative to the plane of the planets orbiting it ("Solar Obliquity Induced by Planet Nine," *Astronomical Journal* 152:156, 2016). This is mind-blowing, changing our entire view of our neighborhood. But no one has yet seen "Planet Nine," or "Planet X," as it is sometimes called.

For current news and images from the exploration of the solar system, go to the NASA website.

MOTHERHOOD

Bishop James Ussher's (1581–1656) dating of the earth has been subject to ridicule in our times, among other places in the Jerome Lawrence and Robert Edwin Lee play *Inherit the Wind*, which premiered in 1955, and the fantasy novel *Good Omens* (Gollancz, 1990) by Terry Pratchett and Neil Gaiman, in which Ussher is accused, tongue-in-cheek, of being "off by a quarter of an hour." But read Stephen Jay Gould's essay "Fall in the House of Ussher," in his book *Eight Little Piggies: Reflections in Natural History* (W. W. Norton, 1993), which takes a different tack, showing Ussher's chronology to be the honorable, and impressive, effort that it was for the times.

For a history of nineteenth- and early-twentieth-century debates about the age of the earth, centered on the geologist Arthur Holmes but including Kelvin, Rutherford, and many others, have a look at Cherry Lewis's *The Dating Game: One Man's Search for the Age of the Earth* (Cambridge University Press, 2000).

For millennia the origins of the moon remained a mystery. It wasn't until the 1940s that Reginald A. Daly suggested that the moon was created by a giant impact on earth, in "Origins of the Moon and Its Topography," *Proceedings of the American Philosophical Society* 90:104–19, 1946. But Daly's idea was more or less ignored until William K. Hartmann and Donald R. Davis revisited the theory in "Satellite-Sized Planetesimals and Lunar Origin," *Icarus* 24:504–14, 1975. This was followed by Alistair G. W. Cameron and William R. Ward's "The Origin of the Moon," *Abstracts of the Lunar and Planetary Science Conference* 7:120–22, 1976. It's mind-blowing to think that the origin of the moon, almost as basic as motherhood, was worked out only about forty years ago: man walked on the moon before he knew where it came from. It is also hard to fathom that the moon used to loom larger in the sky, and that days were once much shorter, and that it is slowly moving away, at the pace of the growth of fingernails. The man responsible for christening the Mars-size protoplanet that collided with Earth, Theia, after the Greek Titan who gave birth to the moon goddess Selene, was the English geochemist Alex Halliday, by the way, in 2000.

For updates on the relationship between Theia, the earth, and the origin of the moon, see Daniel Herwartz et al., "Identification of the Giant Impactor Theia in Lunar Rocks," *Science* 344:1146–50, 2014; and Edward D. Young et al., "Oxygen Isotopic Evidence for Vigorous Mixing During the Moon-Forming Giant Impact," *Science* 351:493–96, 2016.

The giant impact theory remains a consensus view, but there is a problem: If a single planet hit Earth, creating the moon, one would expect the composition of the moon to be about one-fifth Earth-like and about four-fifths Theia-like. And yet the earth and the moon are made up of almost identical matter. This has long perplexed champions of the giant impact theory. An alternative theory, whereby multiple impacts to the earth of small "planetesimals" threw matter into orbit, which gradually clumped together to form "moonlets," which then clumped together to form the moon, has been around for a while and has received a tailwind from a recent study. According to this theory, there were many small moons before they gradually coalesced to form our moon over millions of years. The formation of the moon, in other words, overlapped with a considerable portion of the earth's own growth. If that's true, perhaps motherhood would mean something entirely different. To learn more, see Raluca Rufu, Oded Aharonson, and Hagai B. Perets, "A Multiple-Impact Origin for the Moon," *Nature Geoscience* 10:89–94, 2017. As ever, science is on the move.

On the intriguing possibility that water came to earth from outer space, see Kathrin Altwegg et al., "67P/Churyumov-Gerasimenko, a Jupiter Family Comet with a High D/H Ratio," *Science* 347:1261952, 2015. And for current news and views of the moon, see again NASA's website.

Why the moon is moving away from Earth at a pace of 3.78 cm, or 1.49 inches, each year, has to do with mechanics. While the moon is kept in orbit by the earth's gravitational pull, it too exercises a gravitational force on the earth, which causes the movement of the earth's oceans (and even, by a few centimeters, of its solid mass), creating a tidal bulge. Because of the rotation of the earth, the tidal bulge actually sits a little ahead of the moon, and some of the spinning earth's energy gets transferred to the moon via friction. Thus feeding a small amount of energy into the moon itself, the tidal bulge pushes it into a slightly higher orbit. If a long-living bug lived on the moon, it might experience it like a children's merry-go-round: the faster the merry-go-round spins, the stronger the feeling of being slung outward. But although the acceleration provided by the tides is actually slowing the moon down, it is also slowing the earth's spin down, a phenomenon that will lengthen our days, extremely slowly, and—long into the future—dramatically affect our climate.

IMMORTALITY

Allen Saunders famously lamented that life is what happens to us while we are making other plans, but I prefer a quote attributed to Albert Einstein: "There are only two ways to live your life. One is as though nothing is a miracle. The other is as though everything is a miracle."

This notion sums up well people's reaction when confronted with what the latest science has to say about the perennial question of how to define life. Two worthwhile bookends for our times are the physicist Erwin Schrödinger's classic *What Is Life?* (Cambridge University Press, 1944) on the one hand, and on the other the biologist and historian Michel Morange's *Life Explained* (2003; English translation, Yale University Press, 2008). The former introduced what was, at midcentury, a new, modern conception of life, based on the quantum and thermodynamic physics of "the hereditary substance" (not yet unveiled as DNA), and the latter clarifies, sixty years on, just how difficult defining life remains.

If a definition of life is difficult, figuring out its origins is even more tricky. There are scores of theories on the origin of life, and the controversy is intense. Here I concentrate on only a few of the more extravagant ones.

Consensus largely holds that life began on planet Earth deep beneath the waves, in and around hydrothermal vents. These were, and remain, otherworldly environments: here the driving force for life was sulfur, searing smoke rose up from the seabed like black pillars, and oxygen was nowhere to be found. For technical expositions of this strange environment and how it gave birth to life, read Gunther Wächtershäuser, "Origins of Life: Life As We Don't Know It," *Science* 289:1307–8, and his "On the Chemistry and Evolution of the Pioneer Organism," *Chemistry and Biodiversity* 4:584–602, 2007. Also, William Martin and Michael J. Russell, "On the Origin of Biochemistry at an Alkaline Hydrothermal Vent," *Philosophical Transactions of the Royal Society B: Biological Sciences* 362:1887–1926, 2007. For a more popular piece on the relationship between the origin of life and rocks, see Nathaniel Comfort, "The Primordial Fertility of Rock: The Chemistry of Life Is an Extension of the Chemistry of the Earth," *Nautilus*, December 2016.

Finding evidence for life's beginnings in the depths wouldn't have been possible without technology. On the exploits of *Alvin*, the deep-sea submersible, see Victoria A. Kaharl, *Water Baby: The Story of Alvin* (Oxford University Press, 1990), as well as Robert Kunzig, *The Restless Sea: Exploring the World Beneath the Waves* (W. W. Norton, 1999). There is also the Woods Hole Oceanographic Institution's *Alvin* webpage, featuring an interactive tour of the submersible.

No less amazing than the hydrothermal vent deep-sea theory is the theory that life originated in the clouds, a minority view though vigorously argued. For the original exposition of this idea, see Carl R. Woese (whom we'll meet again soon), "A Proposal Concerning the Origin of Life on the Planet Earth," *Journal of Molecular Evolution* 13:95–101, 1979, and a later endorsement, answering some of the initial criticisms, by Verne R. Oberbeck, John Marshall, and Thomas Shen, "Prebiotic Chemistry in Clouds," *Journal of Molecular Evolution* 32:296–303, 1991.

As one might imagine, the theory of life originating on Mars is even more controversial. See Christopher P. McKay, "An Origin of Life on Mars," *Cold Spring Harbor Perspectives in Biology* 2:a003509, 2010, and Joseph L. Kirschvink and Benjamin P. Weiss, "Mars, Panspermia, and the Origin of Life: Where Did It All Begin?," *Palaeolontologia Electronica* 4:8–15, 2002. To follow the *Curiosity* mission to Mars, now six years and counting since landing on August 6, 2012, go to the NASA website. And to consider a possible earthly origin of life, though more akin to the one proposed for Mars (in a system of volcanic pools and hot springs on land rather than underwater hydrothermal vents), see Martin J. Van Kranendonk, David W. Deamer, and Tara Djokic, "Life Springs," *Scientific American* 317:28–35, August 2017.

Despite the different theories, almost all scientists agree about the following: to cross the threshold and be considered living, chemicals need to be able both to replicate and to metabolize, quite an unlikely affair.

Wherever life began, therefore, it immediately needed to solve a seeming paradox: how to both replicate and metabolize at the same time, often referred to as the chicken-and-egg problem. DNA can replicate but cannot metabolize, so that wouldn't do. Proteins can metabolize, but they can't replicate—a further rut. In the 1980s, the Nobel laureate biologist Walter Gilbert proposed that the paradox could be untangled if the first molecules of inheritance were made from RNA rather than DNA. In his "RNA world hypothesis," self-replicating, metabolizing ribozymes (kinds of RNA) enclosed within a membrane would have gotten life started before DNA came along, acting as both chicken and egg. Read his original review, "Origin of Life: The RNA World," *Nature* 319:618, 1986, as well as a more recent version by another Nobelist, Thomas R. Cech, "The Ribosome Is a Ribozyme," *Science* 289:878–79, 2000. But the RNA world hypothesis isn't airtight, and many disavow it. Recent work focused on the role of epigenetics even points to the possibility that the chicken-and-egg problem isn't really a paradox at all; to learn more, see Yitzhak Pilpel and Oded Rechavi, "The Lamarckian Chicken and the Darwinian Egg," *Biology Direct* 10:34–38, 2015.

There are many theories on the biological origin of life—see Iris Fry's book *The Emergence of Life on Earth: A Historical and Scientific Overview* (Rutgers University Press, 2000) for a history of the subject. The biochemist Nick Lane emphasizes the important role of energy in the origin-of-life story in his book *The Vital Question*, as does, provocatively, the physicist Jeremy L. England in his paper "Statistical Physics of Self-Replication," *Journal of Chemical Physics* 139:121923, 2013—an approach to entropy that many find interesting but as yet unfounded. Scientists argue about how to define the essence of life; one compelling theoretical framework describes the transition from inanimate matter to life as characterized by the acquisition of unlimited hereditary potential: Tibor Gánti, *The Principles of Life*, with a commentary by James Griesemer and Eörs Szathmáry (Oxford University Press, 2003). But heredity may not have been genetic to begin with. One evolutionary scenario, focused on a "garbage bag" of chemicals that predated fully fledged cell-like entities and template replication, can be found in Vera Vasas et al., "Evolution Before Genes," *Biology Direct* 7:1, 2012. These are but a few of the more enticing theories floating around these days. As you might sense, the mystery is far from solved.

Whether wrought on land or on Mars or deep beneath the waters, whether borne first by RNA or protein, via thermodynamics or basic chemistry, when it did finally get started life depended on a little luck: chance events we would ultimately call "mutations." Unplanned and unplanning, fundamentally mysterious, mutations provide much of the variation that makes evolution possible. Without them, life as we know it would not exist. So chance, it turns out, is indispensable.

More than anything, mutations teach us one important lesson: however much we might want it to be otherwise, the future must remain unknown to us in the present. When all is said and done, the future's very existence depends on its unpredictability.

LOVE

If the origins of life remain mysterious, so too are there many theories about the early evolution of our planet's life-forms. Scientists used to believe that early life had a clear genealogy: from two separate domains of life, Bacteria and Eukarya, all the rest of life evolved in two distinct branches. But thanks to seminal work by Carl Woese beginning in the 1960s, a third and ancient domain, Archaea, was discovered, and things grew more complicated. On the origin of cells, see Carl R. Woese, "On the Evolution of Cells," *Proceedings of the National Academy of Sciences* 99:8742–47, 2002; and on the early phylogeny of life, see C. R. Woese and G. E. Fox, "Phylogenetic Structure of the Prokaryotic Domain: The Primary Kingdoms," *Proceedings of the National Academy of Sciences* 74:5088–90, 1977. As it turns out, Eukaryotes, which would eventually evolve into all the complex multicellular life-forms on earth, including humans, are genetically closer to Archaea, and emanated from them, leading some researchers to argue for just two primary domains of life: see Tom A. Williams et al., "An Archaeal Origin of Eukaryotes Supports Only Two Primary Domains of Life," *Nature* 504:231–36, 2013. These are arguments over classification and, to some degree, over language. But what was the relationship between the domains like, all those years ago?

The theory of horizontal gene transfer (HGT) holds that the genealogy of the two, and then three, domains was complicated by the lateral exchange of DNA with one another: life began as a lacework, it seems, not a lineage. Cell membranes, where they existed, had yet to develop defenses and were permeable, and homeostatic function within cells was rather rudimentary. With little to demarcate self from environment, all living things that met simply merged their hereditary materials. As individuality became more pronounced, this began to change. It was then that a pilus evolved, by which early organisms would confer their DNA upon others, including the genetic transcript that would allow the recipient itself to build a pilus. A billion years later, in the line of the Eukaryotes, DNA was sequestered within a nucleus for the first time, changing the way genetic materials were arranged, and then exchanged, between living things, and ushering in chromosomes, meiosis, and gendered sex (see the myth on Death).

The exchange of genetic materials had undergone dramatic evolutions: it began entirely promiscuous, became reciprocally conditional, and finally was dependent on mating, with all its accompanying complications. To learn more about early life's lacework, as well as the continuing role of horizontal transfers in biology, see Maria Boekels Gogarten, J. Peter Gogarten, and Lorraine Olendzenski, eds., *Horizontal Gene Transfer: Genomes in Flux* (Humana Press, 2009). For a shorter read, focusing on the persistence of horizontal gene transfer in eukaryotes, see Patrick J. Keeling and Jeffrey D. Palmer, "Horizontal Gene Transfer in Eukaryotic Evolution," *Nature Reviews Genetics* 9:605–18, 2008. And for the historically and philosophically inclined, see Carl R. Woese's thoughtful article "A New Biology for a New Century," *Microbiology and Molecular Biology Reviews* 68:173–86, 2004.

Debates about the early evolution of life excite great emotion. One man who has tried to quiet the flames with some sober science has been Ford Doolittle. On Doolittle and his pioneering work in molecular phylogeny, see Maureen A. O'Malley, "W. Ford Doolittle: Evolutionary Provocations and a Pluralistic Vision," in Oren Harman and Michael Dietrich, eds., *Dreamers, Visionaries, and Revolutionaries in the Life Sciences* (University of Chicago Press, 2018). For an accessible description of how molecular analysis has changed our view of early evolution, see Doolittle's own popular article, "Uprooting the Tree of Life," *Scientific American* 282:90–95, 2000.

How might horizontal gene transfer have occurred? Early on, before the existence of semipermeable membranes, anything alive would have just bumped into anything else alive and gushed their genetic materials indiscriminately. But then life became more discerning, evolving mechanisms for controlling the transfer of hereditary materials more closely between separate genetic entities. Joshua Lederberg and Edward Tatum received the 1958 Nobel Prize in Physiology or Medicine for discovering the mechanism of transferring DNA through pili (the plural of pilus), known as bacterial conjugation. Read their classic paper, J. Lederberg and E. L. Tatum, "Gene Recombination in *Escherichia Coli*," *Nature* 158:558, 1946, if you want to experience science in the making. Lederberg and Tatum's discovery was a big deal, since it explained how bacteria have sex and, among other things, acquire immunity. As it turns out, not only is immunity transferred by conjugation, but so too are the very instructions to build a pilus. Thus, being able to give at all is dependent on first receiving the gift to give from others.

As its name suggests, bacterial conjugation is the provenance of bacteria, but there is evidence that early in life's history, Archaea, too, engaged in horizontal transfer, their original pili only later evolving into motile flagella. For the argument and evidence, see Sandy Y. M. Ng et al., "Cell Surface Structures of Archaea," *Journal of Bacteriology* 190:6039–47, 2008. On the birth of the nucleus in eukaryotes, including the suggestion that it might have been a virus that got it all started, see Elizabeth Pennisi, "The Birth of the Nucleus," *Science* 305:766–68, 2004.

The Florentine writer, poet, and humanist Giovanni Boccaccio (1313–1375) is best known as the author of *The Decameron*. But search out his lesser-known slim book, *The Most Pleasant and Delectable Questions of Love*, translated by "H. G." in 1566 (Aegypan, 2006)—it is beautiful. Much earlier, the story of Dido and Aeneas had been recounted in Virgil's *Aeneid*, which was made into a musical opera, *Dido and Aeneas*, by the composer Henry Purcell and the librettist Nahum Tate circa 1688. Both inspired this myth.

FREEDOM

The three slogans of the English Socialist Party of Oceania in George Orwell's novel *1984* are: "War is Peace. Freedom is Slavery. Ignorance is Strength." These are the dictates of Big Brother, and while we resist such pronouncements as malicious, nature beckons us to take a second glance.

In the history of our planet, the transformation from a reducing environment to an oxygenated one was momentous, often called the period of the Great Rusting of the Earth. Scientists believe it was the prerequisite for the continued growth of life, providing sustenance to organisms with growing nervous systems and therefore more-complex behaviors. Without more oxygen, life might have stalled at the single-cell level, and all the diversity we know would have never happened. What allowed for the oxygenation was the invention of photosynthesis, the serendipitous result of a chance meeting between a cyanobacterium and a chlorophyll molecule. To learn more about the original role of photosynthesis in sparking this drama, see Roger Buick, "When Did Oxygenic Photosynthesis Evolve?," *Philosophical Transactions of the Royal Society B: Biological Sciences* 363:2731–43, 2008.

But what did the tree of life look like after it began to really grow? We all know that family history can be tricky—every aunt and uncle and grandpa and distant cousin seems to have a different version of events. But what's true for humans is infinitely true for ancient microorganisms: the tracks of time and the vagaries of preservation render reconstruction of genealogy almost hopeless. Still, scientists have found impressive ways around this, and over the past forty years a clearer picture has been emerging, one in which three initial domains—Archaea, Bacteria, and Eukaryotes—take center stage. Surprisingly, as mentioned in the previous illumination, molecular phylogenies based on the analysis of small RNAs have shown that Eukaryota, which is the branch to which we humans belong, is closer to Archaea than to Bacteria (archaea are older, accounting for the surprise). It is likely, though not certain, that the first eukaryotes shared a common ancestor with Archaea, and that the eukaryote branch began to mutate, losing its cell wall and acquiring a nucleus.

But how did eukaryotes turn from single-cell organisms into complex multicellular creatures like humans? The theory of endosymbiosis, also known as symbiogenesis, has emerged in the past few decades to provide a reply and challenge our received view of evolution. According to the theory, eukaryotes first evolved from simple prokaryotes due to the ingestion by an early life-form first of an ancestor of mitochondria, and later of the ancestor of chloroplasts (a second event, which does not appear in the myth for the sake of simplicity). This turned out to be a momentous occurrence in the history of life because energy production is related to surface-area-to-volume ratio, and is dependent on oxygen: bacteria pump protons across their membranes, so their size is limited by geometrical constraints, whereas eukaryotes (including humans), by internalizing energy production in the form of mitochondria or chloroplasts, don't face the same constraints. This was important: almost immediately, eukaryotes had an incentive to grow, since following the logic of economies of scale, bigger size reduces production costs. Thanks to the availability of more oxygen and an in-house system to marshal it for energy, the greatest watersheds of life now came about: organisms grew tens of thousands of times bigger, accumulated thousands of times more DNA, and developed into multicellular forms (including, eventually, humans). Complex life, it seems, could not have assembled without endosymbiosis. By different means, mitochondria and chloroplasts helped achieve similar ends, producing energy-storing molecules (especially ATP) that help their hosts grow. Each is considered the *sine qua non* of all animals and plants, respectively. You might ask: Why were bacteria left behind? Why didn't they too swallow their own in-house energy makers? And why did the specific circumstances that brought about the eukaryotes happen just once? The answer might simply be: chance. This doesn't mean we can't try our hand at ex post facto explanations. See, for example, Marcello Barbieri, "How Did the Eukaryotes Evolve?," *Biological Theory* 12:13–26, 2017.

The theory of endosymbiosis, or symbiogenesis, had antecedents but was vigorously propounded by the biologist Lynn Margulis in the late 1960s and 1970s, and was initially considered fantastic by many. Over time, despite the ruckus, endosymbiosis has been quietly accepted as a crucial factor in evolution. "Lynn could infuriate her colleagues," a fellow scientist remembered in an obituary, "but at least one of her proposals changed the way we think about life."

To read the original paper, see Lynn Sagan (she was married then to the astronomer Carl Sagan), "On the Origin of Mitosing Cells," *Journal of Theoretical Biology* 14:255–74, 1967. For a broader evolutionary exposition, see Lynn Margulis and Dorion Sagan, *Microcosmos: Four Billion Years of Microbial Evolution* (Summit, 1986; reprint ed., University of California Press, 1997), and *Acquiring Genomes: A Theory of the Origin of Species* (Basic, 2002). And on Lynn Margulis's own fascinating life, including her trials and tribulations, see her son Dorian Sagan's *Lynn Margulis: The Life and Legacy of a Scientific Rebel* (Chelsea Green Publishing, 2012).

Molecular phylogenies do not cease to surprise us, and one thing we have discovered is that the harbinger of life and the harbinger of death are closer than we dared to think: the nearest genetic relative of human mitochondria, it transpires, is none other than the bacterium rickettsia, purveyor of typhus. What scientists believe happened in evolution is that the ancestor of human mitochondria and the ancestor of modern rickettsia diverged when the first was unwittingly swallowed, becoming an endosymbiont, while the other retained its freedom, all those years ago. See Michael W. Gray, "Rickettsia, Typhus and the Mitochondrial Connection," *Nature* 396:109–10, 1998, as well as Victor V. Emelyanov, "Evolutionary Relationship of Rickettsiae and Mitochondria," *FEBS Letters* 501:11–18, 2001.

For a short summary of the exploits of typhus, one of the great scourges of humankind, see George Cowan, *The Most Fatal Distemper: Typhus in History* (Lulu.com, 2016). For more specific in-depth histories, see Stephan Talty, *The Illustrious Dead: The Terrifying Story of How Typhus Killed Napoleon's Greatest Army* (Crown, 2009), and Arthur Allen, *The Fantastic Laboratory of Dr. Weigl: How Two Brave Scientists Battled Typhus and Sabotaged the Nazis* (W. W. Norton, 2014). And on the man after whom the typhus-carrying bacterium is named, who both studied the disease and succumbed to it, see Dominik Gross and Gereon Schäfer, "100th Anniversary of the Death of Ricketts: Howard Taylor Ricketts (1871–1910). The Namesake of the Rickettsiaceae Family," *Microbes and Infection* 13:10–13, 2010.

The president who lost his son to typhus was Franklin Pierce, in 1843. On the latest known circumstances of Anne Frank's death, see Michael Winter, "New Research Sets Anne Frank's Death Earlier," *USA Today*, March 31, 2015.

And so, finally, when you contemplate the split between rickettsia and mitochondria all those years ago, remember that life isn't always what we think it is. We may abhor the dictates of Big Brother, but the great drama of evolution, at the very least, poses a challenge to our moral intuitions: "War is Peace. Freedom is Slavery. Ignorance is Strength."

DEATH

It was the husband-and-wife team of Leda Cosmides and John Tooby who argued convincingly that the tiny symbiont mitochondrion is the unlikely culprit in the birth of sex and gender in nature. When asexual reproduction reached an impasse, the energy-giving hitchhiker did what it needed to do to survive. And since two armies of mitochondria were bound to clash in the fusing algae, a division of labor came into effect. See Cosmides and Tooby's original paper, "Cytoplasmic Inheritance and

Intragenomic Conflict," *Journal of Theoretical Biology* 89:83–129, 1981, for details.

As ever in nature, there are exceptions to the rule: the "split gill" mushroom *Schizophyllum commune* has 28,000 mating types, for example. But the deepest difference between the sexes relates to intolerance: the internal organelles of fusing single-cell organisms simply could not countenance each other, and the least complicated solution turned out to be uniparental inheritance of mitochondria (and chloroplasts, which again don't appear in this myth for simplicity's sake). Gradually, the differences sharpened between those who brought the contents of their cytoplasm to the union and those who didn't. That is why, despite the price, almost everywhere in nature we find two genders. To learn more about the paradox of sex and about different theories of its origin, read John Maynard Smith's classic book *The Evolution of Sex* (Cambridge University Press, 1978) and take it from there.

While *Chlamydomonas* may be a killer, silencing its "lover's" mitochondria, for most life-forms that turned to sexuality the solution was simply avoidance: eggs would pass on their mitochondria while sperm contributed nuclear DNA exclusively, thereby subverting battle. The arrangement does not always go over quietly: selfish mitochondrial DNA has found ingenious ways of making it into the next generation despite nature's machinations. Darwin himself wrote about "male cytoplasmic sterility"—how in angiosperms, mitochondria avoid ending up in pollen by sterilizing the male sex organs, converting the hermaphrodite into a female to safeguard their own transmission. But there is also "heteroplasmy," whereby mitochondria from both sexes exist in the germ line—in bats, for example. We humans have taken such mixes to a new level with "ooplasmic transfer," a form of in vitro fertilization in which mitochondrial DNA from a female donor is injected together with sperm DNA into the egg of an infertile woman. For more on this controversial technology, see Jessica Hamzelou, "Everything You Wanted to Know About '3-Parent' Babies," *New Scientist*, September 28, 2016. The salient point is that if the deepest distinction between the sexes relates to restricting germ-line passage of mitochondria to the female, the barrier between the sexes is not as high as we believe.

Mitochondrion got sex and gender rolling by letting out a nervous burst of toxic free radicals emanating from its electron transport chain, but things changed once multicellular life came around. Now a germ line was created, apart from the rest of the body, and the mechanism that formerly compelled single-cell organisms to fuse was retooled to get rid of deviants that didn't chime with the whole organism's reproduction. The housekeeping performed by mitochondria is a form of programmed cell death: between 50 billion and 70 billion cells die each day in adult humans, and that's a normal part of development. In the 1960s it was termed "apoptosis," Greek for the dropping off of petals from flowers, and in 2002 Sydney Brenner, H. Robert Horvitz, and John E. Sulston won the Nobel Prize in Physiology or Medicine for identifying the genes that control the process. For an introduction to the important role of apoptosis in living creatures, see the relevant passages in Bruce Alberts et al., eds., *Molecular Biology of the Cell*, 6th ed. (Garland Science, 2014). On August Weismann, the nineteenth-century biologist and father of the germ plasm theory (also once known as "Weismannism"), see Frederick B. Churchill, "August Weismann and a Break from Tradition," *Journal of the History of Biology* 1:91–112, 1968; Rasmus G. Winther, "August Weismann on Germ-Plasm Variation," *Journal of the History of Biology* 34:517–55, 2001; and Charlotte Weissman, "The First Evolutionary Synthesis: August Weismann and the Origins of Neo-Darwinism," PhD Diss., Tel Aviv University, 2011.

Alas, the ancients knew: Eros is inextricably linked to Thanatos. Thus the story of mitochondria does not end with sex and gender and mere apoptosis, as crucial as this last function is to forging multicellular life and individuality. Even the best machines break down, and life is no exception: ultimately, the mitochondrial process by which the electron transport chain utilizes electrons to oxidize hydronium ions into water begins to leak, and highly reactive free radicals are produced, creating chains of even more free radicals, and wreaking havoc on the body's membranes, proteins, and DNA. On the connection between the gradual winding down of mitochondrial function and senescence, see two classic papers: Denham Harman, "The Biologic Clock: The Mitochondria?," *Journal of the American Geriatrics Society* 20:145–47, 1972; and J. Miquel et al., "Mitochondrial Role in Cell Aging," *Experimental Gerontology* 15:575–91, 1980. For an excellent overview of the important role of mitochondria in evolution, read Nick Lane's *Power, Sex, Suicide: Mitochondria and the Meaning of Life* (Oxford University Press, 2005), and for more-up-to-date data on cell death and senescence, see David M. Hockenbery, ed., *Mitochondria and Cell Death* (Humana Press, 2016).

Interestingly, from the early days of developmental biology, there have always been those who believe that if we just figure out the principles of growth, science will be able to beat back death and promise immortality. For a comprehensive history of this movement, see Ilia Stambler, *A History of Life Extensionism in the Twentieth Century* (CreateSpace Independent Publishing Platform, 2014). For a more popular account, see Jonathan Weiner, *Long for This World: The Strange Science of Immortality* (Ecco, 2010).

Sex and death, it seems, are two sides of a flipping coin: in the end, what brought us here contributes to our demise. Some may view this as a kind of Greek tragedy. For my own part I take the counsel of a wise man, Dr. Seuss: "Don't cry because it's over," he is (perhaps falsely) attributed as saying. "Smile because it happened."

PRIDE

The Princeton microbiologist John Tyler Bonner devoted his life to studying the social amoeba *Dictyostelium discoideum*, often referred to as "Dicty," a modern descendant of the amoebae that first came on land and began to congregate. Bonner shared much of his sixty years of insight on this remarkable organism, and some of its brothers and cousins, in *The Social Amoebae: The Biology of Cellular Slime Molds* (Princeton University Press, 2009). Social amoebae are incredible altruists in the biological sense, "sacrificing" themselves under conditions of duress in such a way that allows some of their kind to keep on living. Remarkably, however, like all social creatures, they harbor free riders too, and have even been shown to have evolved policing. For the original experiments showing cheating in mixed clones of Dicty, see Joan E. Strassmann, Yong Zhu, and David C. Queller, "Altruism and Social Cheating in the Social Amoeba *Dictyostelium discoideum*," *Nature*, 408:965–67, 2000. While behavior is far from genetically determined in humans, at least in relatively simple creatures cheating and cooperation are all in the genes: see Elizabeth A. Ostrowski et al., "Genomic Signatures of Cooperation and Conflict in the Social Amoeba," *Current Biology* 25:1661–65, 2015.

When and how multicellular life first came about remains a mystery, though it is sure to have occurred multiple times and represents a giant leap in the evolution of life. What is clear is that it led, separately, once to animals, once to plants, and on various occasions to fungi, seaweed, ciliates, slime molds, and other small, mainly aquatic creatures. It is likely that multicellular life first appeared not by individual cells coming together but rather by one cell that divided and then failed to separate, more than a billion years ago in the ocean. Those two cells in turn then divided and stuck together too, again and again.

In different lineages, different paths were followed from single-cell individuals to multicell communities. What unites them all, however, is that the autonomy of the individual became subsumed by the collective, even if temporarily, as in the life cycle of cellular slime molds (which probably evolved somewhat later than did the first aquatic multicellulars). To learn more about all these different paths and lineages, see John Tyler Bonner, "The Origins of Multicellularity," *Integrative Biology* 1:27–36, 1998, and the more recent Andrew H. Knoll, "The Multiple Origins of Complex Multicellularity," *Annual Review of Earth and Planetary Sciences* 39:217–39, 2011.

On the enigmatic, self-contained, and pacific biota of the Ediacaran period, 635–542 million years ago, see Mark A. S. McMenamin, *The Garden of Ediacara: Discovering the First Complex Life* (Columbia University Press, 1998), in which Reg Sprigg's exploits are recounted and a strange world seemingly devoid of much interaction between organisms is brought to life. On the Cryogenian, which preceded the Ediacaran and represents the greatest ice age known to have occurred on the planet, search under "The Proterozoic Eon" at Palaeos.com, an offbeat but informative website on "life through deep time." And on Rodinia, the supercontinent that was assembled between 1.3 billion and 900 million years ago and broke up some 250 million to 300 million years later, see S. V. Bogdanova, S. A. Pisarevsky, and Z. X. Li, "Assembly and Breakup of Rodinia (Some Results of IGCP Project 440)," *Stratigraphy and Geological Correlation* 17:259–74, 2009.

Why multicellularity entails cooperation but also necessarily selfishness and deceit has to do with the dynamics of populations under natural selection. While groups competing with one another will tend to become altruistic within themselves, so as to reap the benefits of cooperation, it is almost always in the interest of the individual within the group to shirk the taxes imposed by being part of a community for its own selfish good. There is therefore an inbuilt conflict between different levels of the biological world: genes, chromosomes, cells, individuals, groups, even lineages. To learn more about these dynamics, see the somewhat technical but lucid *Evolution and the Levels of Selection* (Oxford University Press, 2006) by Samir Okasha, as well as his plea "Altruism Researchers Must Cooperate," *Nature* 467:653–55, 2010.

Intriguingly, multicellularity connects disease, altruism, individuality, and pride. On cancer, a disease of the failure of cooperative multicellularity, see Siddhartha Mukherjee, *The Emperor of All Maladies: A Biography of Cancer* (Scribner, 2011). On altruism, including the difference between altruism in humans and other creatures, see Oren Harman, *The Price of Altruism: George Price and the Search for the Origins of Kindness* (W. W. Norton, 2010). On biological individuality, see Leo W. Buss, *The Evolution of Individuality* (Princeton Legacy Library, 1988). On the microbiome, or the fact that we have as many cells in our bodies that are not us as are, see Ed Yong, *I Contain Multitudes: The Microbes Within Us and a Grander View of Life* (Ecco, 2016). And on pride, consider C. S. Lewis: "A proud man is always looking down on things and people: and, of course, as long as you are looking down, you cannot see something that is above you." But pay heed, too, to Voltaire, who perhaps better than anyone captured our multicellular, multisocial conundrum: "We are rarely proud," he wrote, "when we are alone."

JEALOUSY

In the geological record, the Ediacaran is followed by the Cambrian, usually dated from 542 million to 488 million years ago, and in most accounts presented as a true revolution in the history of life. From the outset, the Cambrian was dramatic: Charles Darwin even mentioned in the *Origin* that it represented one of the biggest objections to his gradual theory of evolution. Whereas earlier creatures were comparatively small, few and far between, devoid of armor or weapons, and rather self-contained and uninterested in one another, an abundance of complex, interactive, predacious animals now appeared in the world, suddenly, as if from behind a lifted curtain. In an instant in evolutionary terms, the rate of diversification of species had increased by over an order of magnitude, or so the accounts went. Colorfully described by Stephen Jay Gould in his book *Wonderful Life: The Burgess Shale and the Nature of History* (W. W. Norton, 1989), hard-shelled arthropods such as *Opabinia*, *Hallucigenia*, and *Anomalocaris* now appeared with fully developed nervous systems, segmented limbs, gills, antennae, claws, and even dramatic mastodon-like tusks. Such ferocious animals swam between gentle sea lilies like *Echmatocrinus*, a creature resembling a flower attached to the sea bottom by a stalk and sporting colorful tentacles on its crown. Due to the suddenness with which it appeared and the efflorescence of life it introduced, the period was dubbed the "Cambrian Explosion."

In recent years, this view of life has come up for criticism: Was there really an "explosion" of life in the Cambrian? Many scientists are no longer so sure. Based on new findings from China to Greenland to Namibia, a plethora of complex creatures turn out to have already been present at the start of the Cambrian. And so it seems that Darwin was right all along: life emerged more gradually. As the British paleontologist Richard Fortey writes in his book *Trilobite: Eyewitness to Evolution* (Knopf, 2000): "When a play (particularly a whodunnit) is beginning to stall a little, one standard bit of theatrical 'business' to bring it alive is to introduce an explosion. Boom! The audience jumps to attention; and, of course, under the cover of gun-shot it is possible to get away with murder, dramatically speaking."

Whether or not the word "explosion" has been misused, it remains the case that in the period between roughly 650 million and 500 million years ago, almost all the basic body plans of life we know (and many that we no longer recognize) came about and flourished on Earth. Why this happened has been the subject of intense debate, and in recent years the focus has been on oxygen. The rise in oxygen levels in the Earth's oceans, so the argument goes, allowed organisms, including their nervous systems, to become more complex. See Douglas Fox's short article, "What Sparked the Cambrian Explosion?," *Nature* 530:268–70, 2016. To delve somewhat more deeply into the possible role of the evolving nervous system, read Detlev Arendt, Maria Antonietta Tosches, and Heather Marlow, "From Nerve Net to Nerve Ring, Nerve Cord and Brain—

Evolution of the Nervous System," *Nature Reviews Neuroscience* 17:61–72, 2016.

One of the complex features attributable to the rise in oxygen levels was the eye. And while primitive vision may have started as early as 750 million years ago, the earliest fossilized eyes belonged to a class of 20,000 species of marine arthropods, some as small as mosquitoes and others bigger than giant turtles, all sporting three vertical lobes. These are called trilobites. The only eyes ever to be built from calcite, trilobite eyes were amazing contraptions: Euan N. K. Clarkson and Riccardo Levi-Setti have shown that such eyes anticipated designs offered by the Dutch scientist Christiaan Huygens and French polymath René Descartes in the seventeenth century, as an optical "cure" for spherical aberration ("Trilobite Eyes and the Optics of Des Cartes and Huygens," *Nature* 254:663–67, 1975). As Fortey writes: "This may indeed be a wonderful example of Art imitating Nature, or perhaps rather of Nature anticipating Science—by more than 400 million years." On the wonder of the evolution of vision in nature more generally, look at the colorful and accessible *The Evolution of the Eye* by Georg Glaeser and Hannes F. Paulus (Springer, 2015).

Consisting of rock fitted into a molting carapace, trilobite eyes did not blink: there is even one species named *Opipeuter inconnivus*, Greek and Latin for "one who gazes without sleeping." Constructed of up to thousands of calcite rods, trilobite eyes provided a marvelously sensitive view of their landscape. (Researcher Kenneth Towe of the Smithsonian has even photographed the FBI building across the street from his lab from within their lenses in order to visualize precisely what they saw, but don't tell the authorities!) Certain species of trilobites lost their eyes in the course of evolution, but there are no known examples of eyes that were lost and regained: vision, it would seem, is a one-way street. The "Light Switch" theory of Cambrian diversification speculates that eyes put evolution into overdrive by improving locomotion, hence intensifying predation, hence enhancing competition, hence initiating an arms race. To protect themselves, marine arthropods, including trilobites, developed tough outer skeletons made of calcium carbonate, which guaranteed their excellent fossil record. As the inventions proliferated, so did the number of species, leading to the great diversification of life. And so with help from a little bit of oxygen, eyes forged the living world. For an account of this idea, see Andrew Parker, *In the Blink of an Eye: How Vision Kick-Started the Big Bang of Evolution* (2nd revised ed., The Natural History Museum, 2016).

Eyes may have also introduced behaviors for a different kind of competition. We do not know much about trilobite sexuality, but it looks as if sexual selection was rampant, which means that not all males could always get what they wanted. Evidence of this is presented in Robert J. Knell and Richard A. Fortey, "Trilobite Spines and Beetle Horns: Sexual Selection in the Palaeozoic?," *Biological Letters* 1:196–99, 2005. On the rape of Tamar, daughter of King David, by her half brother Amnon, see 2 Samuel 13; on the rape of Cassandra, daughter of King Priam and Queen Hecuba of Troy, by Ajax the Lesser—two figures who appear in Homer's *Iliad* XXIV and *Odyssey* XI—see the fragments of Alcaeus; and on the rape of Persephone by Hades, see the *Homeric Hymn to Demeter*.

Whether or not competing trilobite males sometimes turned to coercion to survive, eyes—both real and metaphorical—are a mixed blessing.

CURIOSITY

Curiosity killed the cat, goes the saying, but in truth there would be no cat to kill were it not for the impulse to explore new ideas and environments in the first place. The evolutionary process that got us here, as well as the minds that have helped us figure out evolution, owe whatever success they have achieved to curiosity's indelible link to survival.

How life develops from a fertilized egg has been a long-standing mystery; Matthew Cobb's book *Generation: The Seventeenth-Century Scientists Who Unraveled the Secrets of Sex, Life, and Growth* (Bloomsbury, 2006) provides a history of early attempts to solve the conundrum. We have come a long way since the battles between those who believed organisms were perfectly and minutely preformed in the eye of God and simply grew larger during development in the egg or the womb (preformationists), and those who thought life assembles piecemeal—also orchestrated by the deity—in a gradual working out of form (epigenesists). For a collection of essays detailing this history, including discussions of von Baer and Spemann, as well as Roux and His and Morgan and other important embryologists into our times, see *From Embryology to Evo-Devo: A History of Developmental Evolution* (MIT Press, 2007), edited by Manfred D. Laubichler and Jane Maienschein. For more on Ernst Haeckel, his times, philosophy, and biogenetic law ("ontogeny recapitulates phylogeny"), see Robert J. Richards, *The Tragic Sense of Life: Ernst Haeckel and the Struggle over Evolutionary Thought* (University of Chicago Press, 2008). And on Hilde Mangold (and other great women scientists, oftentimes not sufficiently acknowledged), have a look at Marilyn Ogilvie, Joy Harvey, and Margaret Rossiter, eds., *The Biographical Dictionary of Women in Science: Pioneering Lives from Ancient Times to the Mid-20th Century* (2 vols.) (Routledge, 2000).

As it turns out, the preformationists and the epigenesists each held some piece of the truth. For as developmental biologists in the 1980s began working out the genetics of growth and form, it transpired that while life is itself never preformed, there are nevertheless highly conserved genes, called *Hox* genes, which direct the business of development, from sea squirts to humans, and have changed very little in hundreds of millions of years. (*Hox* genes are so named because the early-twentieth-century English geneticist William Bateson called oddities such as legs growing where eyes should form in flies "homeobox mutations." For more on Bateson, see Patrick Bateson, "William Bateson: A Biologist Ahead of His Time," *Journal of Genetics* 81:49–58, 2002.) Form is a delicate assemblage—a symphony of timing—but its instructions are abiding. Thus, the business of embryogenesis is an amalgam of ancient and new, preformed and emergent. To learn more about where mainstream science stands today, read Sean B. Carroll's accessible book *Endless Forms Most Beautiful: The New Science of Evo Devo* (W. W. Norton, 2005), and see Scott F. Gilbert and Michael J. F. Barresi's excellent textbook *Developmental Biology*, 11th ed. (Sinauer/Oxford University Press, 2016).

Genes, however, may not always be the leaders in evolution. In fact, they may actually be the followers. Development itself is a flexible, exploratory process, and an organism's physiological and behavioral adaptations to new environments in many cases precede the genes that stabilize changes to its behavior and form. To dive deeper into the ways in which this emerging understanding is changing our view of evolution, including by blurring the traditional lines between organism and environment as well as between germ line and soma, read *Evolution:*

The Extended Synthesis (MIT Press, 2010), edited by Massimo Pigliucci and Gerd B. Müller, as well as a more recent thematic issue of the Royal Society journal *Interface Focus* (7:20170051, 2017) titled "New Trends in Evolutionary Biology: Biological, Philosophical and Social Science Perspectives," organized by Denis Noble, Nancy Cartwright, Patrick Bateson, John Dupré, and Kevin Laland. Have a look, too, at Denis Noble's wonderfully lyrical book, *The Music of Life: Biology Beyond Genes* (Oxford University Press, 2006). And for an even more heretical suggestion, see Oren Harman, "In the Future There Will Be No Genes," *Acta Biotheoretica*, 66:2, 2018.

How life first came out of the oceans and onto land is one of evolution's great stories, featuring correlated progression, changing environments, mutating genes, mutualist bugs, and serendipity. While somewhat dated, the story is told well in Carl Zimmer's *At the Water's Edge: Fish with Fingers, Whales with Legs, and How Life Came Ashore but Then Went Back to Sea* (Touchstone, 1998). Some years after Zimmer's book was published, *Tiktaalik* was discovered in the Canadian arctic by the University of Chicago paleontologist and developmental biologist Neil Shubin, as detailed in his book *Your Inner Fish: A Journey into the 3.5-Billion-Year History of the Human Body* (Pantheon, 2008). *Tiktaalik* lacked fingers, but it had a shoulder, with a humerus, radius, ulna, and wrist bones. Its fins were distinctly limb-like, and seem to have specialized to push its head and upper body above oxygen-depleted shallows in ancient brackish inlets. This was an organism that was almost ready to move from water to land, a harbinger of the transformation of fish into tetrapod amniotes—our direct ancestors. Shubin's genetic work uncovering the *Hox* genes responsible for limb development has been reported by Zimmer: "From Fins into Hands: Scientists Discover a Deep Evolutionary Link," *New York Times*, August 17, 2016.

Whether or not *Tiktaalik* is the last word in the transition from fish to land-dwelling amphibians and reptiles, the precise circumstances of the birth of the first tetrapods and their emergence onto land remain elusive. It used to be believed that limbs and lungs evolved from the necessity of having to find new bodies of water as old ones dried up—this was called the "shrinking water hole hypothesis" or the "desert hypothesis." But this theory withered away when more careful analysis showed that most of the innovations that describe the evolution from fish to amphibians and amniotes—such as the birth of lungs and a neck and a shoulder, elbow, and wrist, as well as the accompanying loss of buoyancy and incipient creation of the ear bones—all happened first in the waters and only later took on new functions on land, a phenomenon known as exaptation. Today, two theories continue to compete for supremacy: based on the discovery of 395-million-year-old tracks in Zachelmie, Poland, the "intertidal hypothesis" argues that some lobe-fish, or sarcopterygians, came onto land from the area that is above water at low tide and underwater at high tide. The "woodland hypothesis," on the other hand, bases its claims on the fact that tetrapod fossils have been consistently found in habitats that were once humid and forested. Navigating shallow waters filled with roots and vegetation would have made limbs and fingers extremely useful in such habitats, creating the selection pressures that brought about the creatures that first came onto land.

As new fossils are discovered, the story will be reshaped and retold; a coda has yet to be written. Still, it is worth remembering that wisdom always comes from ignorance, which curiosity entails. Undoubtedly, our wisdom will be scored ignorance by our successors. As long as we continue to be curious.

SOLITUDE

Prematurity is perhaps the fiercest form of solitude. Just ask the lizards. Or anyone who has ever had an idea that came before its time.

It is true: flight was invented by the insects. Scientists argue over a number of scenarios, including that wings first evolved from gills among insects' marine crustaceous ancestors, in the waters, or that they arrived on land, emanating from the thorax of early terrestrial hexapods. To learn more about these debates, read Ivar Hasenfuss, "The Evolutionary Pathway to Insect Flight—a Tentative Reconstruction," *Arthropod Systematics and Phylogeny* 66:19–35, 2008; and the informative online discussion by Marc Srour, "Insect Flight: Origins and Aerodynamics," at Bioteaching.com.

The far greater challenge of getting vertebrates up in the air stemmed from the fact that almost all the lift to counteract drag would need to be used to carry the weight rather than motor the thrust of any heavenly pretender, a plea that might seem audacious to gravity. As it turned out, the pterosaurs were the first to crack it, 220 million years ago, though they would not survive to brag about it. On the amazing pterosaurs, including *Quetzalcoatlus northropi*, see Mark P. Witton's book *Pterosaurs: Natural History, Evolution, Anatomy* (Princeton University Press, 2013). For gorgeous visualizations, available in 3-D, see David Attenborough's 2011 BBC film, *Flying Monsters.*

As for the story of the dramatic discovery of *Quetzalcoatlus northropi*, in his own words, read Douglas A. Lawson, "Pterosaur from the Latest Cretaceous of West Texas. Discovery of the Largest Flying Creature," *Science* 187:947–48, 1975; Crawford H. Greenewalt's questioning "Could Pterosaurs Fly?," *Science* 188:676, 1975; and Lawson's response immediately following (676–78). When *Quetzalcoatlus northropi* was introduced to the world it was a slow news day and the *San Angelo Times* changed its lead from "Man Trapped in New York Subway" to "Largest Flying Creature Discovered in Texas." For an entertaining retrospective of the event, see Stephen Harrigan, "The Miracle of Flight," *Alcalde*, Nov–Dec 2013.

It is likely that pterosaurs disappeared for the same reason as the dinosaurs: none have been found in the fossil record after the asteroid that created the Chicxulub crater in the Yucatán impacted, marking the Cretaceous-Tertiary boundary 66 million years ago. It may be, however, that the fate of the pterosaurs was sealed long before this: birds' rigid flight feathers meant their wings had no need to be anchored to either flanks or legs, allowing them to run and walk and pounce freely, whatever they needed to survive and flourish. (For the argument that birds were in fact dinosaurs, see Gregory S. Paul's beautifully illustrated *Dinosaurs of the Air: The Evolution and Loss of Flight in Dinosaurs and Birds* [Johns Hopkins University Press, 2002]). Encumbered by their skinny wings, pterosaurs might have already been at a disadvantage before the asteroid hit, though looking at *Quetzalcoatlus* bones, one wonders. In the end, these can only be conjectures. The disappearance of the pterosaurs remains a mystery.

So, too, does the decision of the U.S. Air Force to scrap the B-49 (and B-35) prototype remain a puzzle. Already, Northrop's brainchild had outperformed the competition, but post–World War II big business intervened, and politics, and perhaps jealousy and avarice. It is never easy, and extremely lonely, to be a pioneer. For a glimpse into the story, watch the original documentary film *Flying Wings: John K. Northrop's Final*

Interview—1979, available on YouTube, where you can hear the details from the mouth of Northrop himself, his eyes vacant and his face gaunt.

John Knudsen "Jack" Northrop led a fascinating life. To learn more about him and the flying wing, read Ted Coleman's *Jack Northrop and the Flying Wing: The Real Story Behind the Stealth Bomber* (Paragon, 1988), and E. T. Wooldridge's *Winged Wonders: The Story of the Flying Wings* (Smithsonian Institution Press, 1983).

SACRIFICE

Aristotle didn't quite know how to score them. He thought them a separate group, like insects or birds. It wasn't until the seventeenth century that scientists concluded: whales must be mammals, an unthinkable truth. If anything, cetaceans have taught us this signal lesson: evolution not only creates anatomy but can sometimes dramatically erase it.

Melville writes in *Moby-Dick*: "Consider the subtleness of the sea; how its most dreaded creatures glide under water, unapparent for the most part, and treacherously hidden beneath the loveliest tints of azure." And indeed whales have been feared for most of human history. This is a mark of our ignorance, principally: leagues more about these creatures remain a mystery than is known. "Yea, foolish mortals," we're reminded, "Noah's flood is not yet subsided; two thirds of the fair world it yet covers."

Darwin wrote in the first editions of the *Origin*: "I can see no difficulty in a race of bears being rendered, by natural selection, more and more aquatic in their structure and habits, with larger and larger mouths, till a creature was produced as monstrous as a whale." He was on the right track. But it was the anatomist and surgeon William Henry Flower (1831–1899), director of the Natural History Museum in London, who showed that whales had evolved from terrestrial mammals. The names people called them—sea hogs, sea pigs, herring hogs, porcpoisson—weren't far off after all: the ancestor had not been a bear but was most probably a land-trotting ungulate, perhaps an ancient hippopotamus. Flower divided cetaceans into two separate groups: mysticetes, with baleen in their mouths, like the giant blue whale or smaller minke—the aquatic filter feeders; and the toothed whales, or odontocetes, including killer whales, sperm whales, dolphins, porpoises, and narwhals. These split from each other, we now think, some 34 million years ago, but all have blowholes, and flippers attached to shoulders, and unlike the side-to-side fishes, they swim by bending their back up and down with the help of former tails that are now flukes. For a rhapsodic cultural meditation on whales, wrapped in a literary history of the writing of Melville's *Moby-Dick*, treat yourselves to Philip Hoare's award-winning book, *The Whale: In Search of the Giants of the Sea* (Ecco, 2010). For a modern account of the transformation of whales in human eyes from grotesque monsters to playful friends, see D. Graham Burnett, *The Sounding of the Whale: Science and Cetaceans in the Twentieth Century* (University of Chicago Press, 2012).

Whale evolution is a tale yet unfolding, but we know much more today than we did in Flower's day. After it was established, primarily by Leigh Van Valen in the 1960s and based mainly on fossil teeth, that whales were formerly ungulates rather than carnivorous dogs—in technical terms, mesonychid descendants of condylarths rather than creodonts—Phillip Gingerich went out to look for fossil remains along the shores of the former Tethys Sea, in latter-day Pakistan. In a series of dramatic discoveries, he uncovered first *Pakicetus*, the Pakistani whale, some 50 million years old, a terrestrial animal that foraged in streams, six feet long and looking a bit like a deer or coyote. Then came *Ambulocetus*, the walking whale, a few million years younger, fourteen feet long and feeding in the waters, resembling an alligator. Fast-forward a few million more, and *Maiacetus* makes its entrance, already aquatic, with nostrils climbing up its snout and shortened forelimbs, although—like a sea lion—spending time on land. Thirty-seven million years ago, finally, *Dorudon* joins the older, lizard-like *Basilosaurus*, both with flukes, streamlined, their habits and physiology entirely aquatic. It was perhaps the starkest sequence of evolution in action yet. And when among the *Basilosaurus* vertebrae Gingerich discovered a knee and ankle and three perfect toes, little doubt remained as to the whales' origin. For an account of the discoveries by a paleontologist who himself played an important role in the drama (among other things discovering *Ambulocetus*), see J. G. M. "Hans" Thewissen, *The Walking Whales: From Land to Water in Eight Million Years* (University of California Press, 2014).

There are in fact many more fossil species in the whale evolution sequence, including *Kutchicetus* and *Rodhocetus*. Increasingly, science is figuring out the genetic and environmental causes for their dramatic transformations (and has recently challenged the mesonychid origin, offering an artiodactyla one instead). *Hox* genes, unsurprisingly, play a crucial role, but so do geological conditions, the ebbing of seas and the birth of mountains and aquatic sanctuaries. Science is also uncovering the mysteries of the evolution of echolocation, a technology hit upon by early toothed whales following the split with the baleen filterers (only toothed whales echolocate, though baleen whales originally had teeth). Figuring out how echolocation evolved is difficult: most of the implicated parts—the muscles and melons and lips through which the sound squeezes—are soft and do not preserve. Still, skulls carry important clues. For a study charting the possible birth of echolocation in the extinct Oligocene whale *Cotylocara macei*, a beautifully preserved fossil skull of which was recently found in a drainage ditch in North Carolina, see Jonathan H. Geisler, Mathew W. Colbert, and James L. Carew, "A New Fossil Species Supports an Early Origin for Toothed Whale Echolocation," *Nature* 508:383–86, 2014. *Cotylocara*, by the way, diverged from an ancestor of filter-feeding whales, and left no direct lineage. It was a second echolocating branch from which all living toothed whales emerged.

Much remains mysterious about cetaceans. The ability to hunt together, including by intricately coordinated bubble blowing (see the gorgeous footage of Alaskan humpbacks fishing in this manner in the BBC documentary series *Nature's Great Events*), to communicate via song and clicks, even to use tools and instruct calves, all undoubtedly attest to unusual intelligence. So does research in captivity pointing to astonishing feats of memory and language and recognition of self. But we know very little of the inner worlds of cetaceans. Milton wrote in *Paradise Lost*: "There Leviathan / Hugest of living creatures, on the deep / Stretch'd like a promontory sleeps or swims / And seems a moving land; and at his gills / Draws in, and at his breath spouts out a sea." Indeed, whales are an awesome sight. To begin to touch the mystery, read Hal Whitehead and Luke Rendell's book *The Cultural Lives of Whales and Dolphins* (University of Chicago Press, 2014). But to really sense the magic, jump into their waters.

MEMORY

Consciousness can be thought of as akin to the origin of life: How did matter transition from something inert into something living, that could replicate and metabolize? Here too there must have been some sort of a transition, from organisms that only sensed to organisms that experienced—those for whom it suddenly felt like something to be themselves, almost miraculously.

The philosopher Thomas Nagel famously asked, "What is it like to be a bat?" (*Philosophical Review* 83:435–50, 1974). His answer was frustrating: We will never know. Ludwig Wittgenstein agreed. If lions could speak, he wrote in his *Philosophical Investigations*, we would not understand them. But to appreciate the mystery of consciousness, do we really have to turn to lions and bats? We humans want to believe that language can bridge the gap and get us into someone else's brain, but can we ever really comprehend what it even feels like to be another person?

There will probably always be those who view the problem of consciousness as intractable, and perhaps they are not mistaken: it may be true that the nature of consciousness is in the mind of the beholder rather than the eye of the observer. Read the neurologist Robert A. Burton's book *A Skeptic's Guide to the Mind: What Neuroscience Can and Cannot Tell Us About Ourselves* (St. Martin's, 2013) for a wise and witty exposition of this view, and Sally Satel and Scott O. Lilienfeld's *Brainwashed: The Seductive Appeal of Mindless Neuroscience* (Basic, 2013) for a dose of sober skepticism on the entire enterprise.

But there are also those who believe consciousness is explainable. The philosopher Galen Strawson recently presented the thesis in "Consciousness Isn't a Mystery. It's Matter," in the column *The Stone* in the opinion section of *The New York Times* (May 16, 2016). Strawson quotes Bertrand Russell in a public talk in 1950 titled "Mind and Matter," delivered only hours after he'd discovered he'd been awarded the Nobel Prize: "We know nothing about the intrinsic quality of physical events except when these are mental events that we directly experience." Russell was turning the problem on its head: one needn't become a dualist, or otherwise deny the very existence of experience; mind and brain are merely the same thing looked upon from different angles (psychology and physics). We all know what it's like to experience the coldness of a frog (an example provided by Russell), or the taste of a pineapple (Locke), or the sensation of hearing jazz (the philosopher Ned Block via Louis Armstrong)—that's no puzzle. But such knowledge is the only insight we have into the intrinsic nature of reality. For matter, not consciousness, is the true conundrum.

This, of course, doesn't mean that matter will be able to understand itself, but there are people out there trying. For a current rendering of Russell's same idea but with a personal twist, read *Consciousness: Confessions of a Romantic Reductionist* (MIT Press, 2012) by the neuroscientist Christof Koch, president and chief scientific officer of the Allen Institute for Brain Science in Seattle. Together with Francis Crick, Koch described what the two called "the neural correlates of consciousness" in a path-breaking paper nearly thirty years ago. To get a feel for where cutting-edge empirical and modeling research on consciousness stands today, see Stanislas Dehaene, *Consciousness and the Brain: Deciphering How the Brain Codes Our Thoughts* (Viking Penguin, 2014), as well as Michael S. Gazzaniga's recent book *The Consciousness Instinct: Unraveling the Mystery of How the Brain Makes the Mind* (Farrar, Straus and Giroux, 2018). And to better understand the notion of a "remembered present," which draws both from the neuroanatomist Santiago Ramón y Cajal and the philosopher William James, placing consciousness squarely in a Darwinian framework, see Gerald M. Edelman, *The Remembered Present: A Biological Theory of Consciousness* (Basic, 1989).

Recently, the theoretical biologist Eva Jablonka and the physiologist Simona Ginsburg have been attempting to present an evolutionary framework for the birth of consciousness; according to them, consciousness can be better understood if it is thought of as an evolved thing, something that came about by degrees rather than fully formed, beginning hundreds of millions of years ago. Their journey has taken them back in time to ancient comb jellies, or cnidarians, who could sense but not yet experience. These simple organisms, and then the jellyfish-like ctenophores (and acoelomorphs—perhaps our direct ancestors), had diffuse nervous systems that allowed them to react reflexively to their environment, though they did not yet have a brain. Gradually, from the "white noise" of internal sensing, selective stabilization of the neural network brought about simple learning via stimulus and response—a "remembered present." Departures from homeostasis could now produce "primordial emotions," and modifications of reflex, for some things were more or less pleasant. The next stage had to do with the decoupling of stimulus from response, which ushered in associative learning—a huge leap forward. For the ability to retain memory traces of previously selectively stabilized neural connections even when the stimulus was no longer present opened up a "remembered future," and in effect brought about a past. This is when fully fledged feelings evolved, and the compulsion of wanting something—a brave new world had begun.

To learn more about these exciting new ideas, see Simona Ginsburg and Eva Jablonka, "The Teleological Transitions in Evolution: A Gántian View," *Journal of Theoretical Biology* 381:55–60, 2015, which argues that a method for describing markers in the evolutionary transition from nonliving to living can be applied to the evolution of consciousness, as well as to rationality and moral reasoning; Zohar Z. Bronfman, Simona Ginsburg, and Eva Jablonka, "The Transition to Minimal Consciousness Through the Evolution of Associative Learning," *Frontiers in Psychology* 7:1954, 2016, which delves more deeply into the specific biology; and Simona Ginsburg and Eva Jablonka, "Experiencing: A Jamesian Approach," *Journal of Consciousness Studies* 17:102–24, 2010, which ties their evolutionary scenario, as Edelman did his, to William James's philosophy. Look out for Jablonka and Ginsburg's book on the subject, forthcoming. In the meantime, to contrast the two competing views of physicalism and dualism in its modern guise, respectively, compare Daniel C. Dennett's *From Bacteria to Bach and Back: The Evolution of Minds* (W. W. Norton, 2017) to David Chalmers's *The Character of Consciousness* (Oxford University Press, 2010). How we understand consciousness has an impact on how we think about agency, and vice versa. To gain insight into what is at stake in the argument between the "representation" versus "embodied" approaches to cognition, as it relates to both natural and artificial intelligence, read the MIT roboticist Rodney A. Brooks's manifesto "Elephants Don't Play Chess," *Robotics and Autonomous Systems* 6:3–15, 1990. Abstract representationalists focus on the brain as a computational machine, and embodied emergentists on cognition as occurring in a body experiencing its environment over time, but both share a common conviction: the old Cartesian assumption that a material being can have no inner rational agent, no unified self. Agency is only apparent. To consider a counterargument to both approaches, see Thomas Nagel's own *Mind and Cosmos: Why the Materialist Neo-Darwinian Conception of Nature Is Almost Certainly Wrong* (Oxford University Press, 2012), in which he argues, controversially, that our current accounts of evolution, much less our attempts at AI, are not adequate to explain the phenomena of purposeful life.

Wherever you end up standing on such matters, neurons probably first evolved sometime before jellyfish and other cnidarians, perhaps among the sponges. When oxygen levels rose to a certain level, organisms began to grow larger and live longer, and a new body plan was introduced—the bilaterians, with a head and tail and back and belly. Around that time, some 600 million years ago, a momentous split occurred in evolution: down one line came all the vertebrates, and down the other came all the mollusks.

Clams, oysters, snails, and worms are all mollusks, but so are cephalopods: octopuses, cuttlefish, squid, and the equally otherworldly nautilus. While the nautilus remained in its shell, and therefore is comparatively dull-witted, the others, the octopus in particular, continue to astonish humans with their intelligence. As the philosopher Peter Godfrey-Smith explains, there have been two grand experiments in the evolution of mind on our planet, of which we humans and the cephalopods are representative. While we have central nervous systems, cephalopods evolved "distributed intelligence": their arms are suffused with neurons and are semiautonomous, meaning that their mind is literally spread throughout their bodies. These two very different neural architectures, we imagine, combined with the fact that we are social creatures while most cephalopods are not, produced extremely different ways of experiencing the world, and likely a different consciousness. Meeting an octopus underwater really is the closest we can ever come to meeting an alien.

Here are some facts about cephalopods: Their eyes have no blind spot. They have three hearts and blue-green blood (since their blood carries copper, not iron). Their tantalizing camouflage and communication abilities are based on multilayered screen-like skins harboring chromatophores, iridophores (modulated by the neurotransmitter acetylcholine, implicated in human memory, arousal, and motivation), and leucophores. They evolved from a limpet-like Ediacaran seafloor dweller, well before the dinosaurs.

Here are some specific facts about octopuses: They have eight arms, which can continue to function when severed from their bodies. They recognize individual people, and are known to harbor strong likes and dislikes. They are extremely inquisitive and also extremely unpredictable. They live short lives: most species one or two years at most. An octopus with a six-yard arm span, weighing one hundred pounds, can fit into an opening an inch wide, about the size of its eyeball. Octopuses in captivity are inveterate escape artists, and have been known to short-circuit electricity supply by squirting water at lightbulbs, and to flood lab floors by plugging valves in their tanks. Certain species can mimic up to fifteen different organisms, including flounder, lionfish, and sea snakes, just by changing shape and color. Octopuses are color-blind. They are worshipped in Shinto shrines. Some have a detachable penis. Linnaeus dubbed them *Singulare monstrum*, or "unique monster."

To think deeply about what it might mean to feel like one, and why, see Peter Godfrey-Smith's beautiful book *Other Minds: The Octopus, the Sea, and the Deep Origins of Consciousness* (Farrar, Straus and Giroux, 2016). Another illuminating book, by researchers who have focused on octopus memory and its relation to consciousness, is Jennifer A. Mather, Roland C. Anderson, and James B. Wood, *Octopus: The Ocean's Intelligent Invertebrate* (Timber Press, 2010). Some readers may also enjoy the popular account by the science writer Sy Montgomery, *The Soul of an Octopus: A Surprising Exploration into the Wonder of Consciousness* (Atria, 2015). Fascinating work on the relationship between the octopus brain and body is being carried out in the lab of Benny Hochner at the Hebrew University. Read Guy Levi and Binyamin Hochner, "Embodied Organization of *Octopus vulgaris* Morphology, Vision, and Locomotion," *Frontiers in Physiology* 8:164, 2017, to get a taste of the action.

Will we ever understand consciousness? I do not know. Albert Einstein insisted that no problem can be solved from the same level of consciousness that created it, and they don't call him Einstein for nothing. The great Argentinian writer Jorge Luis Borges, for his part, explored the relationship between memory and consciousness in his short story from 1942, "Funes the Memorious." And the composer Charles Ives said, "A rare experience of a moment at daybreak, when something in nature seems to reveal all consciousness, cannot be explained at noon. Yet it is part of the day's unity." All three, alongside memorable underwater encounters with cephalopods, inspired this myth.

TRUTH

Some believe that truth came from God, and the word was with God. But there are others who think God came from falsehood, and falsehood from truth, and that truth was invented by the word, which according to John 1:1 came in the beginning. As Yogi Berra said, "If you don't know where you are going, you might end up someplace else." Mind-bending and mysterious are the ways of evolution.

"Speak and I shall baptize thee!" said the Cardinal of Polignac to a young chimpanzee in the Jardin du Roi in the eighteenth century, or so, at least, claimed Diderot. For generations it was believed that language is what sets humans apart, the *sine qua non* of our intelligence. When Noam Chomsky turned intuition into science in the 1950s, the linguistic revolution had arrived. All of Chomsky's constructs—the critical period of acquisition, grammar universals, overgeneralization of rules, recursion, and the apparent robustness of developmental stages—pointed in one direction: we were *Homo linguisticus*, uniquely, and it was this attribute that explained our great leap forward.

With time it became apparent that there is abundant communication in all of nature's kingdoms, from apes to apple trees, lizards to amoebae. Humans may be the only species with speech, but everyone else is also talking to each other. For an overview of the animal literature, see Jack W. Bradbury and Sandra L. Vehrencamp's textbook *Principles of Animal Communication*, 2nd ed. (Sinauer/Oxford University Press, 2011), and for an entertaining book on plants, read Daniel Chamovitz, *What a Plant Knows: A Field Guide to the Senses* (Scientific American/Farrar, Straus and Giroux, 2012), as well as Peter Wohlleben's *The Hidden Life of Trees: What They Feel, How They Communicate—Discoveries from a Secret World* (Greystone Books, 2016). Even bacteria engage regularly in signaling: educate yourself with *Chemical Communication Among Bacteria*, Stephen C. Winans and Bonnie L. Bassler, eds. (ASM Press, 2008). Still, for more than one reason, we are right to think that there is something special about our human language.

The relationship between thought and language is a fascinating one, and one that we still know little about. Have a look at Michael Tomasello's excellent *A Natural History of Human Thinking* (Harvard University Press, 2014) to learn more about this, but also consider the writer

Cormac McCarthy's take, "The Kekulé Problem: Where Did Language Come From?," *Nautilus*, April 20, 2017—it provides a thought-provoking insight into the relationship between the prelinguistic, unconscious brain and the linguistic, conscious one. What we are starting to learn is that whether or not *Homo sapiens* is a direct descendant of *Homo erectus*, *erectus* was beginning to think abstractly—see, for example, Ewen Callaway, "*Homo erectus* Made World's Oldest Doodle 500,000 Years Ago," *Nature*, December 3, 2014; and Kim Sterelny's excellent *The Evolved Apprentice: How Evolution Made Humans Unique* (MIT Press, 2012). There is reason to believe that short of full-blown speech, *erectus* probably already possessed the rudiments of language, and most assuredly mimesis, the culturally significant ability to imitate. Richard Wrangham, a leading primatologist, has argued forcefully that taming fire nearly 2 million years ago played a major role in this and other hominin developments. See his *Catching Fire: How Cooking Made Us Human* (Basic, 2009). Many anthropologists disagree with Wrangham's thesis, arguing instead that prior to the advent of cooking, our ancestors began eating meats, which caused the shift to smaller guts and larger brains. Whether fire was what made us human or not, there is broad agreement that cooperative social interaction is the key to our cognition. In order to survive, early hominids needed to be able to see the world from different perspectives, get into each other's minds, and develop what has been called a "shared intentionality." Looking into each other's eyes, mimesis, and finally language all played important roles.

But was Chomsky right? Did language only become possible when a specific "language organ" developed in the brain, together with other physiological architectures, and was it an inborn instinct rather than an acquired social tool? For many years most cognitive scientists, linguists, and developmental psychologists believed this to be true, but the paradigm is under attack. Tomasello's own book *Constructing a Language: A Usage-Based Theory of Language Acquisition* (Harvard University Press, 2003) provides examples of exceptions, arguing that language is an ability interwoven into our general cognitive abilities, not a unique blade on the brain's Swiss Army knife, and that it is acquired socially, not inherited biologically. The missionary-turned-linguist Daniel Everett sparked enormous controversy when he claimed that the Pirahã people of the Amazon do not speak with recursion (though they do think recursively), an exception that threatened to topple the entire Chomskyan edifice (see his "Chomsky, Wolfe, and Me," *Aeon*, January 10, 2017). Daniel Dor has recently offered a comprehensive analysis of language as a social communication technology, one that, beginning with *Homo erectus*, was collectively designed for the instruction of the imagination. He argues that our language-ready brains and physiologies were forced into existence by language, not the other way around. See *The Instruction of Imagination: Language as a Social Communication Technology* (Oxford University Press, 2015).

However language got started, it was adapted to convey information, and that information could be either true or false. If language was to survive, so went the argument, lying would need to be constrained significantly, for otherwise the whole communication system would implode. Thus were many arguments presented for reining in falsehood: Tomasello offered the evolution of morality as an antidote in *A Natural History of Human Morality* (Harvard University Press, 2016); Peter J. Richerson and Robert Boyd offered enforcement of norms through punishment in *Not by Genes Alone: How Culture Transformed Human Evolution* (University of Chicago Press, 2005); W. Tecumseh Fitch argued that language evolved first among kin, which ensured its honesty, in *The Evolution of Language* (Cambridge University Press, 2010); Robin I. M. Dunbar suggested that gossip evolved to safeguard against free-riding in "Gossip in Evolutionary Perspective," *Review of General Psychology* 8:100–110; and Chris Knight put forward the theory that individual deception was overcome by the invention of the collective lie in "Ritual/Speech Coevolution: A Solution to the Problem of Deception," in James R. Hurford, Michael Studdert-Kennedy, and Chris Knight, eds., *Approaches to the Evolution of Language* (Cambridge University Press, 1998), 68–91.

But what if the lie was as important to hominin evolution as truthfulness? What if, in fact, language provided great opportunity for forms of falsehood, and that these played a crucial role in allowing *Homo erectus*, followed by *Homo sapiens*, to instruct each other's imagination? Hominin cognition was forged in the crucible of the arms race between liars and lie detectors, giving us imagination. Fiction, on this account, would have played a central role in turning us into humans, gradually widening the gap between us and other primates in our lineage. This is the argument made suggestively by Dor in "The Role of the Lie in the

Evolution of Human Language," *Language Sciences* 63:44–59, 2017. There is also Robert Trivers's entertaining and controversial book *The Folly of Fools: The Logic of Deceit and Self-Deception in Human Life* (Basic, 2011), which argues that in order to best deceive others we must first deceive ourselves.

HOPE

Every person has a path. Mine began in certain ways with Brenda Ralph Lewis's *The World of Myth and Legend* (Brimax Books, 1980) and the stories of Icarus and Theseus, Sir Gawain and Beowulf, Jason and the Golden Fleece, Achilles and Hector, Thor and the giant Rungnir, Odysseus and the Cyclops, Geirrodur the Troll King, and many more.

Early on, alongside the lure of mythology, it was evolution that caught my interest, though I couldn't fathom it at the start. Before I could fully grasp them, I loved reading Stephen Jay Gould's learned essays in *Natural History* magazine, with their colorful characters, human and otherwise, and their historical and zoological arcana. Finally, at university, when Harvey Lodish et al.'s textbook *Molecular Cell Biology* and Eric Kandel et al.'s *Principles of Neural Science* sparked a thirst for context, I read *On the Origin of Species* for the first time, and slowly the insight grew in me. Soon I tore through *The Voyage of the Beagle* and the *Autobiography*, *The Descent of Man*, *The Expression of the Emotions in Man and Animal*, *The Variation of Animals and Plants Under Domestication*, even *The Formation of Vegetable Mould Through the Action of Worms*. I just couldn't get enough of Charles Darwin, and I knew now, undoubtedly: all of life had descended with modification.

I could feel my eyes open when I read Bertrand Russell's *History of Western Philosophy* (Allen and Unwin, 1946), with its lucid biographical snapshots and pithy turns of phrase. The wonder deepened as I experienced the philosophers in their own words in the beautiful 1952 olive and mauve and lazuline leather-bound series of Great Books from Britannica. I followed my reading of Wittgenstein's seminal *Tractatus Logico-Philosophicus* with Ray Monk's brilliant biography, *Ludwig Wittgenstein: The Duty of Genius* (Free Press, 1990).

Later I turned to the history of science, starting with George Sarton's

A History of Science: Ancient Science Through the Golden Age of Greece (Harvard University Press, 1952; repr. Dover, 1993), and his admonition (elsewhere) that "above all, we must celebrate heroism when we come across it." Gradually, positivism gave way to a more sober attitude. Working my way through the writings of the masters, G. E. R. Lloyd and Alexandre Koyré, I. Bernard Cohen and Richard S. Westfall, I came like many to Thomas Kuhn, who taught me that Aristotelian physics was entirely mistaken, though it had captured the minds of the world's smartest thinkers, and had worked for nearly two thousand years. Science, it turns out, has a politics, and its practitioners have been all too human. This insight was taken up by the radical sociologists of the Edinburgh and Bath Schools, David Bloor and Harry Collins, Don MacKenzie, and Steven Shapin. Both "true" and "false" scientific theories, they argued, are a form of social relations, just like literature and art. Whether one believed such heresies or not, it was clear that what counts as "scientific" has never been universal. As Lorraine J. Daston and Peter Galison showed beautifully in their book *Objectivity* (The MIT Press, 2007), science's epistemic virtues were always in flux.

But hadn't the march of reason made the world a better place? In many ways it undoubtedly had. For one thing, we were killing each other less—see Steven Pinker's *The Better Angels of Our Nature: Why Violence Has Declined* (Viking, 2011) for the dramatic evidence. We were living longer and eating better, and mind-numbing, backbreaking human labor was being gradually replaced by machines. More than ever before, people were being educated, traveling, opening up their minds to new ideas based on evidence, and participating in the political process that impacts their lives. Surely, the ideal of the Enlightenment was praiseworthy, and the advances it had set in motion astonishing—for a flawed defense, see Steven Pinker once more, *Enlightenment Now: The Case for Reason, Science, Humanism, and Progress* (Viking, 2018). Still, the Enlightenment's celebration of Reason and Progress was not an unqualified triumph. Maybe Bruno Latour was right that *We Have Never Been Modern* (Harvard University Press, 1993). Certainly, as Ulrich Beck argued in *Risk Society* (Sage, 1992), modernity had failed to deliver on many of its promises. Perhaps, as the late historian-philosopher Yehuda Elkana suggested, it is time for "Rethinking—Not Unthinking—the Enlightenment" (in Wilhelm Krull, ed., *Debates on Issues of Our Common Future* [Velbrück

Wissenschaft, 2000], 283–313). Perhaps we might keep the baby that Karl Popper called the method of *Conjectures and Refutations* (Routledge, 1963) while throwing out the bathwater of dogmatic rationalism, dogmatic objectivity, dogmatic belief in value-free social science, methodological individualism, biological reductionism, and a context-independent moral universalism. Maybe then we'd be able to understand the world more profoundly, and make it a better place.

For me, the way forward seems first to point back into our deep evolutionary past. Where we came from and how we evolved will not determine our fate, but these are important questions with important consequences. A contemporary voice on such matters is the anthropologist Christopher Boehm, who in *Moral Origins: The Evolution of Virtue, Altruism and Shame* (Basic, 2012) shows why we remain today so tribal, punitive, gossipy, religious, and cooperative. While biological-cultural coevolution remains poorly understood, a lot of what Boehm describes can be placed within a framework suggested by the evolutionary biologist Stephen Stearns in his article "Are We Stalled Part Way Through a Major Evolutionary Transition from Individual to Group?," *Evolution* 61:2275–80, 2007. Stearns argues that we may share much of the division of labor, cooperation, and altruism that characterizes the truly social insects, but that we retain individual freedoms and ambitions that they don't possess. This is a good thing, and it makes all the difference, but it also carries a price. For an argument about why ants and humans have become so successful, read E. O. Wilson, *The Social Conquest of Earth* (Liveright, 2012), and a short critique of the book in Oren Harman, "Shakespeare Among the Ants," *Studies in History and Philosophy of Biological and Biomedical Sciences* 44:114–18, 2013. And for an evolution-informed philosophical consideration of both the role of emotions in the birth of morals and the continued importance of culture in human evolution, see Jesse Prinz, *The Emotional Construction of Morals* (Oxford University Press, 2008) and *Beyond Human Nature: How Culture and Experience Shape the Human Mind* (W. W. Norton, 2014).

A further line of inquiry, related to but distinct from the one examining the consequences of our moral natures, has been the study of our intelligence. It was Amos Tversky and Daniel Kahneman who first showed how humans depart from rationality in systematic ways—read their classic paper "Judgment Under Uncertainty: Heuristics and Biases," *Science*

185:1124–31, 1974, to return to the moment, and Michael Lewis's double biography, *The Undoing Project: A Friendship That Changed Our Minds* (W. W. Norton, 2016), for the greater story. Kahneman himself summed up many of the subsequent insights of prospect theory in a wonderful popular book, *Thinking, Fast and Slow* (Farrar, Straus and Giroux, 2011). Interestingly, it turns out that many of the classic heuristics and biases were already known 2,500 years ago, to Plato. See Nick Romeo, "Platonically Irrational," *Aeon*, May 15, 2017.

That we can imagine infinities while being permanently transfixed in finite bodies is our lot in life, metaphorically speaking. For a beautiful tale of men who tried to break the boundary, treat yourself to Loren Graham and Jean-Michel Kantor's book *Naming Infinity: A True Story of Religious Mysticism and Mathematical Creativity* (Belknap, 2009). And for one of the most poignant accounts of the waning of the powers of intellect, read G. H. Hardy's *A Mathematician's Apology* (Cambridge University Press, 1940, repr. 1967). The eighteenth-century materialist Julien Offray de la Mettrie, reader to King Frederick II of Prussia, never believed that man was built to know infinity. "Man in his first principle is nothing but a Worm," he wrote in his *L'homme machine* ("Man a Machine") from 1747, a book that was banned in his day and led to a frenzied escape from France. "Let us not lose ourselves in the infinite, we are not made to have the least idea of it; it is absolutely impossible for us to go back to the origin of things." The theologians and Cartesian philosophers may have believed that reason could extricate man from his material body, but this was a bloated fantasy, and they—vain pretenders. To embrace the ignorance of a fundamentally material creature—a worm, a mole, a crawling machine—was to live a good and just and happy life. This meant connecting to our feelings and emotions and experiences. And accepting, ultimately, that there is much that exists beyond our grasp.

Today we know: there are patterns to world mythologies that can be studied like the patterns of Earth's geology. The myths of different cultures relate to one another because they have a history and have migrated with peoples and have been locally adapted and transformed; see Julien d'Huy, "The Evolution of Myths," *Scientific American* 315:56–63, December 2016. Myths, and storytelling more generally, may have played a crucial role in rendering early groups of humans more cooperative; see, for example, Daniel Smith et al., "Cooperation and the Evolution of Hunter-Gatherer Storytelling," *Nature Communications* 8:1853, 2017. There are also attempts to use artifacts, including myths, to penetrate the (far from primitive) minds of ancient peoples. Start with Alexander Marshack's *The Roots of Civilization: The Cognitive Beginnings of Man's First Art, Symbol, and Notation* (1971; McGraw-Hill First Edition hardcover, 1991), and take it from there.

And yet a science of mythology will not explain away mythology. Despite the fascinating insights we are gaining from a rational look at our legends, we intuit correctly that the great questions dealt by the myths remain unsolved. In some deep way, the more we learn, the more the myths become mysterious. Perhaps this is how it should be. And perhaps it could never be any other way.

ACKNOWLEDGMENTS

The idea for this book has been in my heart since I was a boy, but began in earnest fifteen years ago in a canoe on the Zambezi River with Samantha Power. I thank her for her enthusiasm and sweet encouragment. I also want to thank my good friend Hugh Nissenson—a true artist, who did not live to see *Evolutions*, but who inspired me in many ways—as well as Marilyn Nissenson, for her continued friendship.

Ofra Kobliner drew the wonderful illustrations for the myths, and working with her was a special experience. Thank you, Ofra, for your artistry and quiet wisdom. Thanks to Peter Godfrey-Smith and to the late Jack Repcheck, with whom I shared my developing thoughts. Janet Browne was my generous host at the History of Science Department at Harvard University, where I began writing the book whilst on sabbatical. Thank you, Janet, and thank you, Walden Pond and New England foliage and all our friends, for a magical year.

When I met Eric Chinski at the old FSG offices at 18 West Eighteenth Street, I knew I would do this book with him, though I'm not sure he did at first. What a fortuitous meeting it was! I could not have hoped for a more intelligent partner in bringing the myths to readers. Thank you, Eric, for your gentle and stellar judgment, and for believing in the project when it was still unknown to itself. A great big thank you, too, to Julia Ringo for all her assistance, Jeff Seroy and Stephen Weil and the publicity team, Jonathan D. Lippincott for his marvelous book design, Rodrigo Corral and Sungpyo Hong for their jacket design, and Annie Gottlieb for her incredible Mantis-shrimp eyes.

Thanks to my agent, Sarah Chalfant, for her unwavering support and her brilliance, as well as to Luke Ingram and the rest of the Wiley team. And thanks to my friend Doron Weber, and the Alfred P. Sloan Foundation, who generously supported this project.

My mother and father introduced me to myths when I was a child, to the American Museum of Natural History, and to many other things besides. I love you with all my heart, and you, too, Danz and Mish, my best friends in the universe. Thanks to the Organism, and the Shamna, for all the laughter, and to maestro Michael Shani and my fellow singers at the Tel Aviv Chamber Choir for the music. Finally, to Yaeli, wisest of all women, who read each chapter as it was being born, and who gave us Shaizee and Abie, our most precious treasures: Thank you, my love, for helping us live life as if it were a miracle.